Experimental investigation on turbulent transport in Taylor-Couette flow

Experimental investigation on turbulent transport in Taylor-Couette flow

BRANDENBURGISCH TECHNISCHE UNIVERSITÄT
COTTBUS-SENFTENBERG

DOCTORAL THESIS

Experimental investigation on turbulent transport in Taylor-Couette flow

---

A thesis submitted in fulfilment of the requirements
for the degree of Doctor of Philosophy

in the

Department of Aerodynamics and Fluid Mechanics
Brandenburgisch Technische Universität Cottbus-Senftenberg

by:
Dipl. Phys. Sebastian MERBOLD
born at 10.06.1985 in Erfurt, Germany

June 2019

1. Reviewer:	*Prof. Dr.-Ing.*	*Christoph Egbers*
2. Reviewer:	*Prof. Dr.*	*Innocent Mutabazi*
Chairman:	*Prof. Dr.-Ing.*	*Heiko Schmidt*

Bibliografische Information der Deutschen Nationalbibliothek

Die Deutsche Nationalbibliothek verzeichnet diese Publikation in der Deutschen Nationalbibliografie; detaillierte bibliografische Daten sind im Internet über http://dnb.d-nb.de abrufbar.

1. Aufl. - Göttingen: Cuvillier, 2018

Zugl.: (BTU) Cottbus-Senftenberg, Univ., Diss., 2019

© CUVILLIER VERLAG, Göttingen 2018

Nonnenstieg 8, 37075 Göttingen

Telefon: 0551-54724-0

Telefax: 0551-54724-21

www.cuvillier.de

1. Auflage, 2018

Gedruckt auf umweltfreundlichem, säurefreiem Papier aus nachhaltiger Forstwirtschaft

ISBN 978-3-7369-7068-7

eISBN 978-3-7369-6068-8

"We have just seen that the complexities of things can easily and dramatically escape the simplicity of equations wich describe them. Unaware of the scope of simple equations, man has often concluded that nothing short of God, not mere equations, is required to explain the complexities of the world. "

"The next great era o awakening of human intellect may well produce a method of understanding the qualitative content of equations. Today we cannot. Today we cannot see that the water flow equations contains such things as the barber pole structure of turbulence that one sees between rotating cylinders. Today we cannot see whether Schrödinger's equation contains frogs, musical composers, or morality - or wether it does not. We cannot say whether something beyound it like God is needed, or not. And so we can all hold strong opinions either way "

Richard P. Feynman

The Feynman lectures on Physics (V2 Chapter 41-6 'Couette flow')

Abstract

"Experimental investigation on turbulent transport in Taylor-Couette flow"

A famous example of flow instability occurs between two concentric cylinders, rotating independently. In case of differential rotation angular momentum is transported between the cylinders. The so called Taylor-Couette (TC) flow is of fundamental interest in terms of transport properties in rotating systems, as Earth's atmosphere or accretion disks in case of Rayleigh stable flows or industrial applications such as journal bearings, seperation and mixing for Rayleigh instable flows. A better understanding of the transport processes and the turbulent structures and a uniform description of these complex aspects, these are tasks of great importance for many research areas. The aim of this thesis is to experimentally investigate the turbulent Taylor-Couette flow for radius ratio $\eta = 0.5$ at high shear Reynolds numbers and rotation of inner as well as outer cylinder.

For the present investigation fundamental scientific questions were raised on the topic of turbulent Taylor-Couette flow. *How large is the angular momentum flux for given flow conditions? Does the angular momentum flux scale with Reynolds number comparable with other canonical flows? Do Taylor-vortices exist for large Reynolds numbers? How does the counter-rotation affects the turbulent Taylor-Couette flow?* The investigations on the flow revealed interesting behaviour and opened even more questions: *What drives the torque to be maximal for counter rotation? Which flow structures dominate the counter rotation? Can the torque be predicted by a formula?*

Two experimental setups were realized allowing a precise torque measurement inside the inner cylinder and Laser-optical measurement techniques, while the systems high rotation rates achieve turbulent flows in the classical and ultimate turbulent regime. The angular momentum flux is precisely measured by torque measurements, its scaling and dependencies are determined and discussed, an empirical prediction is revealed. For counter rotation the transport increases, which is against classical expectations. The turbulent Taylor-vortices disappear in the turbulent flow for the ultimate turbulent regime, when the outer cylinder is at rest. For the case of counter rotation the large-scale circulation (LSC) is given the freedom to enhance in the gap. The flow measurements reveal that in these cases the transport process is dominated by the LSC while the turbulent transport decreases - the enlargement of the LSC leads to a total increase of the transport. The measurements are in good agreement with numerical studies and verify theoretical predictions on the transport by the enhanced LSC.

Zusammenfassung

"Experimentelle Untersuchung des turbulenten Transports in der Taylor-Couette Strömung"

Ein berühmtes Beispiel für Strömungsinstabilitäten kommt zwischen zwei konzentrischen, unabhängig rotierenden Zylindern vor. Drehimpuls wird im Falle einer differentiellen Rotation zwischen den Zylindern transportiert. Die so genannte Taylor-Couette (TC) Strömung ist von grundlegender Bedeutung für Transportprozesse in rotierenden Systemen, wie der Erdatmosphäre und Akkretionsscheiben oder industriellen Anwendungen, wie Gleitlagerung, Seperation oder Mischungen. Ein besseres Verständnis der Transportprozesse und turbulenten Strukturen sowie eine einheitliche Beschreibung der komplexen Aspekte sind Aufgaben großer Tragweite vieler Forschungsgebiete. Die Absicht dieser Arbeit ist die experimentelle Untersuchung der turbulenten Taylor-Couette Strömung des Radienverhältnissees $\eta = 0.5$ bei hohen Scher-Reynoldszahlen und der Rotation des Innen- sowie Außenzylinders.

Der vorliegenden Studie wurden grundsätzliche wissenschaftliche Fragen zur turbulenten Tayor-Couette Strömung gestellt. *Wie hoch ist der Drehimpulstransport bei bestimmten Strömungsverhältnissen? Skaliert der Drehimpulstransport mit der Reynoldszahl vergleichbar mit anderen kanonischen Strömungen? Existieren die Taylor-Wirbel bei hohen Reynoldszahlen? Wie beeinflusst die Gegenrotation die turbulente TC Strömung?* Die Untersuchungen haben interessante Strömungsverhalten offengelegt und weitere offene Fragen aufgeworfen: *Was treibt das Maximum des Drehimpulstransports bei Gegenrotation an? Welche Strömungsstrukturen dominieren die Gegenrotation? Kann der Drehimpulstransport vorhergesagt werden?*

Zwei experimentelle Aufbauten wurden realisiert, die eine präzise Bestimmung des Drehmoments im Innenzylinder sowie Laser-optische Messungen erlauben, während die hohen Rotationszahlen Strömung im klassischen und ultimativen turbulenten Regime erlauben. Der Drehimpulstransport wird präzise durch die Drehmomentmesseinrichtung gemessen, die Skalierung und Abhängigkeiten bestimmt und diskutiert sowie eine empirische Vorhersage aufgedeckt. Entgegen klassischen Vermutungen steigt der Transport bei Gegenrotation. Wenn der Außenzylinder steht, verschwinden die Taylor-Wirbel in der turbulenten Strömung im ultimativen turbulenten Regime. Bei Gegenrotation wird der großskaligen Zirkulation (LSC) die Freiheit gegeben sich im Spalt auszubreiten. Die Strömungsmessungen zeigen auf, dass dann die Transportprozesse durch die LSC dominiert werden, während der turbulente Transport sinkt - die Ausdehnung der LSC führt zu einem steigendem Gesamttransport. Die Messungen sind in hervorragender Übereinstimmung mit numerischen Untersuchungen und bestätigen theoretische Vorhersagen des Transports durch die verstärkte LSC.

Acknowledgements

First of all I would like to thank my supervisor Prof. Christoph Egbers at Brandenburg University of Technology Cottbus-Senftenberg for his guidance, the warm environment and familiar climate. You gave me freedom for my scientific development and let me freely navigate through the world of experimental fluid mechanics. I had the chance to work with experts on the state-of-the-art of Turbulence, fluid mechanics, experimental methods and measurement techniques available in the community and gained lots of experience. I would like to thank Prof. Innocent Mutabazi for the fruitful and exciting collaborations we had and the willingness of being my second referee. Also I like to thank Prof. Bruno Eckhardt and Prof. Detlef Lohse on one hand for the interesting theoretical background given to Taylor-Couette flow but also for the great discussions and raising directing questions.

I'd like to thank my past and present colleagues in the Department of Aerodynamics and Fluid Mechanics for sharing good and bad times and for the help on many things; Uwe, Birgit, Nicoletta, Eckhardt, Falko, Yongtai, Paul, Torsten, Franziska, Norman, Sandy, Andreas K., Andreas S, Vasyl, Robin, Ludwig, Florian P., Florian Z., El-Sayed, Gabriel, Marcel, Zeinab, Muhamed, Peter, Markus, Antoine, Martin, Costanza, Wenchao, Vadim and anybody else if I forgot. I really enjoyed my time in the department. I explicitly want to thank my "Taylor-Couette-mate" Andreas Froitzheim, as we worked so well together and shared the passion for this crazy flow between rotating cylinders. Andreas, Gazi Hasanuzzaman and Stefan Richter, you all became good friends to me during the time and I'm happy to know you and I'm proud to work with you. Even if we lost the track i'd like to thank Andreas Christl for the funny times in the department. I would like to appreciate Hannes Brauckmann for the long lasting communication on our shared topic and the great scientific output we got. Also many thanks to Sander Huisman, Roland van der Veen and Rodolfo Ostilla-Monico from the Twente group for the good collaborations we had in lab and conferences. Also great thanks to Borja Arias-Martinez for the share of our interesting field of study. I'd like to thank Norman Jurtz and Steffen Fischer for the great support in designing, engineering and setting up two unique experimental systems, Eduard Pascu for the great assistance in the PIV campaign as well as a great group of different students I was able to advice.

Ein Lehrstuhl braucht auch die Verwaltung. Ich möchte Silke Kaschwich großen Dank aussprechen für all ihre Mühen und Geduld mit uns Wissenschaftlern. Auch braucht ein Experimentator immer wieder hilfreiche Hände, ohne die ein Experiment nie gelingen würde. Vielen Dank an Heinz-Jörg Wengler für seine tatkräftige Hilfe und immerwährende Bereitschaft ein Provisiorium zu finden, großer Dank gebührt auch Hans-Jürgen Pflanz für Präzision, Ausdauer und Witz.

My dear best friend Amir Shahirpour, in any kind of topic - science, music, food, free-
time, live, deep privacy and many more - it is allways a pleasure to share my mind with
you. Thanks again for being there when I experienced the hardest days.

Dank gebührt meinen liebenden Eltern Bärbel und Thomas Merbold, die mir trotz
Schwierigkeiten es ermöglicht haben meinen Weg des Studiums zu gehen und ich mir
Ihrer vollen Unterstützung jeder Zeit bewusst sein konnte. Der gleiche Dank gebührt
meinen Kindern Vincent-Leander und Ella-Jane die mir mit ihrem Lachen so viel Energie
schenken und großer Dank an Rahel für die Unterstützung und Liebe die du mir gibst.

...

Contents

List of Figures

List of Tables

In favour of my children

Vincent Leander

and

Ella Jane.

Chapter 1

Introduction

1.1 Turbulence

The turbulent motion of fuids is still an unsolved problem, while it is a widely distributed appaerence dominating the practical fluid mechanics. The importance of understanding the motion called turbulence follows from its presence in a lot of cases of moving gases and liquids, one can say the majority of all real flows [34]. In engineering one has primarily the flow inside and around technical apparatusus in mind. These flows can be distinguished into a few fundamental types:

- Turbulence while flow through grids (wind tunnel turbulence)

- Flows in pathways (pipes, ducts, channels)

- Free turbulence (Jets, wake flows, areas of mixing)

- Near wall flows (Boundary layers)

- Apart from that the most geophysical flows occur to be turbulent. Those are atmospheric flow, flows in rivers, tides and many more. Also the flow in a climatised room behaves turbulent.

- Also in astrophysics turbulent flows are observable in the chromosphere and star atmosphere while it also plays a crucial role for the development of spiral galaxies and the processes in accretion disks.

The turbulent motion appeals in versatile manners. For example the viscous drag acting on surfaces overflown is much higher for turbulent compared to laminar flow, and it follows different physical laws. On the other hand turbulence can suppress an early separation of the boundary layer, resulting in a reduction of the friction. A fundamental characteristic of turbulent flows is the intense mixing of liquids and/or gases in order that impurities (i.e. dye in water, smoke in air) are rapidly distributed across the volume, mass and heat transfer is huge for turbulent flows. Turbulent flows can cause vibrations at solid body's like wind around buildings. The gust load for air vehicles results from the atmospheric turbulence. Acoustic sound waves passing through turbulent flow experience a scattering and the pressure fluctuations in turbulent flows of high velocities can represent strong acoustic sources, being a serious task for aviation and vehicle aerodynamics.

To give a concisely definition for the understanding what turbulent flow is, is far from simple. The following characteristics can be used to depict turbulent flows:

- 1. *Turbulent flows proceed irregular and unsteady,* such as the velocity variates complicated with space and time. Thus, a single measurement can not deliver a reproducable result, but provide a statistical fluke.

- 2. *Turbulent flows are vortex dominated*

- 3. *Turbulent flows are three dimensional*

- 4. *Turbulent flows are instationary*

The properties given in 3 and 4 are resulting from the irregularity given in 1. Flows not representable by these characteristics are not considered to be turbulent. For example the so called van Kármán street, the flow behind a cylindrical body, is vortex dominated but due to its regularity one does not count it as a turbulent flow. Furthermore the vortex flows generated by the seperation of stream lines at sharp edges are not generally turbulent. At the other hand it exists areas near turbulent flows experiencing velocity fluctuations excited from the pressure fluctuations of the turbulent flow, which are eddy-free and thus, these areas are understood to be non turbulent. As an example for turbulent flow in Figure 1.1 an image of the turbulent atmosphere of the Jupiter is depicted.

A liquid or gas, seen as a continuum, is a mechanical system of infinite degree of freedom. The motions of the fluid particles at different locations in space and time are coupled by the differential equations of fluids, but, in consequence of instability a random encounter of apparently not related circumstances under certain conditions results in an infinite diversity of motions. Such processes where the velocities are random functions of space and time are called stochastic processes. Thus, turbulent flows are understood to be stochastical processes.

As turbulent flows are vortex dominated, the rotation deviates from Zero ($\nabla \times u \neq 0$), it results that viscous forces are not negligible.

FIGURE 1.1: Turbulence in Atmospheres: Jupiter's southern hemisphere in detail, image taken by NASA's Juno spacecraft. The colour-enhanced view captures one of the white ovals in the "String of Pearls," one of eight massive rotating storms at 40 degrees south latitude on the gas giant planet. The image was taken on Oct. 24, 2017 at 11:11 a.m. PDT (2:11 p.m. EDT), as Juno performed its ninth close flyby of Jupiter. The spatial scale in this image is $22.3km/px$. Citizen scientists Gerald Eichstädt and Seán Doran processed this image using data from the JunoCam imager. Image Credits: NASA/JPL-Caltech/SwRI/MSSS/Gerald Eichstädt/ Seán Doran, Source: $https : //www.nasa.gov/mission_pages/juno/images/index.html$, at 24.10.18

A famous example of flow instability occurs between two concentric cylinders of radii R_1, R_2 and length L, rotating independently with the angular velocities Ω_1 and Ω_2. In case of differential rotation angular momentum is transported between the cylinders. The so called Taylor-Couette (TC) flow is of fundamental interest in terms of transport properties in rotating systems, as Earth's atmosphere or accretion disks in case of Rayleigh stable flows or industrial applications such as journal bearings for Rayleigh instable flows. A better understanding of the transport processes and the turbulent structures and a uniform description of these complex aspects, these are tasks of great importance for many research areas van Gils et al. [38]. The aim of this thesis is to experimentally investigate the turbulent Taylor-Couette flow for radius ratio $\eta = 0.5$ at high shear Reynolds numbers and rotation of inner as well as outer cylinder.

For the present investigation fundamental scientific questions were raised on the topic of turbulent Taylor-Couette flow.

- *How large is the angular momentum flux for given flow conditions?*

- *Does the angular momentum flux scale with Reynolds number comparable with other canonical flows?*

- *Do Taylor-vortices exist for large Reynolds numbers?*

- *How does the counter-rotation affects the turbulent Taylor-Couette flow?*

The scope of the experimental study is also limited. The main limitation is the given radius ratio $\eta = 0.5$, which holds for the whole investigation. In parallel also investigations for other radius ratios were performed for more narrow gap: $\eta = 0.71$ investigations were made but have been excluded from this thesis. For wider gaps $\eta < 0.5$ the flow is also analysed and will be the scope of future studies. The shear Reynolds numbers were limited by the maximal driving velocities and reached up to $Re_S < 1.3 \cdot 10^6$. Also the investigation had the scope on centrifugally instable flows, so the behaviour in the Rayleigh-stable regime is not included in this thesis.

1.2 Outline of the thesis

The outline of the thesis is as follows. Chapter 2 explains the physical background on turbulent Taylor-Couette flow which is important for the understanding of the study. Chapter 3 gives details of the two Taylor-Couette experiments built up for this study and explains the principles of the measurement techniques used. In one of the experiments an accurate torque measurement unit was specially designed and implemented. The results of the torque measurements are discussed in Chapter 4.

Chapter 5 consists of velocity measurements by the use of Laser Doppler Velocimetry (LDV) and Particle Image Velocimetry (PIV) revealing statistical velocity profiles in the system. In Chapter 6 a detailed flow visualisation follows. Here the flow behaviour in the concentric gap was analysed qualitively and gives a brief understanding of what kind of structures happen in the Taylor-Couette system and what will be needed for further investigations. The further investigation, in Chapter 7, is directly attached to the flow visualisation. By the use of the same equipment a Particle Image Velocimetry is performed for the flow close to the outer cylinders wall. Here the angular and axial velocities are measured in a short azimuthal and long axial area of interest.

Chapter 8 finally reports a Particle Image Velocimetry at different horizontal planes. From these measurements the three dimensional meanflow can be analysed. The measurements finally give the ability to understand the contributions of the different flow structures onto the angular momentum flux. Chapter 9 finally concludes all over the thesis.

Chapter 2

Theoretical background and theoretical approach

To describe physical system one needs to derive the underlying laws of conservation. In our case we want to investigate the flow of a fluid in rotation. The important laws of conservation in this case are the conservation of mass and momentum. The influence of heat is negligible in this case, so one does need to care about the conservation of energy. In the following chapter the conservation of mass and momentum are used in cylindrical coordinates, using this the Continuity equation and the Navier-Stokes-equation are derived for the Taylor-Couette system.

2.1 Fluid in motion

To describe a fluid in motion in mathematical way one needs a fundamental set of equation. For incompressible fluids, such as we look at the system variables are the velocity field and the pressure distribution. This can be solved taking two conservations laws, the conservation of mass and the conservation of momentum. For fluids these equations become special names: The continuity equation:

$$\frac{\partial \rho}{\partial t} + \nabla \cdot (\rho \vec{u}) = 0 \tag{2.1}$$

with the density ρ, and velocity u. For incompressible fluid $\rho = const.$ it becomes:

$$\nabla \cdot \vec{u} = 0 \tag{2.2}$$

Formulating the momentum balance for fluid flows the Navier-Stokes equation can be written as:

$$\frac{\partial \vec{u}}{\partial t} + (\vec{u}\nabla)\vec{u} = -\frac{1}{\rho}\nabla p + \nu\nabla^2\vec{u} + \vec{f} \tag{2.3}$$

with the pressure p, kinematic viscosity $\nu = \frac{\mu}{\rho}$, dynamic viscosity μ and probable external force fields $\vec{f}$. As one is interested in rotating flows between concentric cylinders the cylindrical coordinates are used. The Nabla and Laplace operator are written as follows.

$$\nabla = \frac{\partial}{\partial r} + \frac{1}{r}\frac{\partial}{\partial \varphi} + \frac{\partial}{\partial z} \tag{2.4}$$

$$\nabla^2 = \frac{\partial^2}{\partial r^2} + \frac{1}{r}\frac{\partial}{\partial r} + \frac{1}{r^2}\frac{\partial^2}{\partial \varphi^2} + \frac{\partial^2}{\partial z^2} \tag{2.5}$$

Thus, the Navier-Stokes equations of an incompressible fluid in cylindrical coordinates reads as:

$$\frac{\partial u_r}{\partial t} = -(\vec{u}\cdot\nabla)u_r + \frac{u_\varphi^2}{r} - \frac{1}{\rho}\frac{\partial p}{\partial r} + \nu\left(\nabla^2 u_r - \frac{u_r}{r^2} - \frac{2}{r^2}\frac{\partial u_\varphi}{\partial \varphi}\right), \tag{2.6}$$

$$\frac{\partial u_\varphi}{\partial t} = -(\vec{u}\cdot\nabla)u_\varphi + \frac{u_r u_\varphi}{r} - \frac{1}{\rho}\frac{\partial p}{\partial \varphi} + \nu\left(\nabla^2 u_\varphi - \frac{u_\varphi}{r^2} - \frac{2}{r^2}\frac{\partial u_r}{\partial \varphi}\right), \tag{2.7}$$

$$\frac{\partial u_z}{\partial t} = -(\vec{u}\cdot\nabla)u_z - \frac{1}{\rho}\frac{\partial p}{\partial z} + \nu\nabla^2 u_z \tag{2.8}$$

The differential operators in these equations on a scalar field Φ are given as:

$$(\vec{u}\cdot\nabla)\Phi = \left(u_r\frac{\partial}{\partial r} + \frac{u_\varphi}{r}\frac{\partial}{\partial \varphi} + u_z\frac{\partial}{\partial z}\right)\Phi \tag{2.9}$$

External force fields can be neglected in our case. If one wants to write the Navier-Stokes equation 2.10 in dimensionless form (all dimensionless variables are noted by index u_* it becomes:

$$\frac{\partial \vec{u}_*}{\partial t_*} + (\vec{u}_*\nabla_*)\vec{u}_* = -\frac{1}{\rho}\nabla_* p_* + \frac{\nu}{LU}\nabla_*^2\vec{u}_* + \vec{f}_* \tag{2.10}$$

were L is a characteristic length and U a characteristic velocity. The dimensionless parameter $Re := \frac{UL}{\nu}$ is called the Reynolds number.

2.2　Taylor-Couette flow

The geometry of the Taylor-Couette system is simple. Two concentric cylinders of Radii $R_{1,2}$ rotate with angular velocities Ω_1, Ω_2, where the indices 1 and 2 denote the inner and outer cylinder respectively, are the boundaries of the geometry. In respect to this a key parameter is the radius ratio $\eta = R_1/R_2$, while the gap width $d = R_2 - R_1$ is a characteristic scale of the system and been used for calculating dimensionless properties.

In a theoretical approach the system has infinite length. In Reality the system has to have axial borders in the form of end plates. The length between the end plates L can be measured in gap widths.

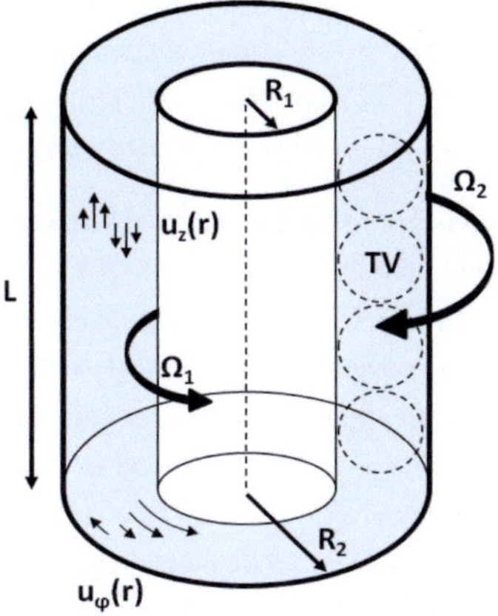

FIGURE 2.1: Sketch of the Taylor-Couette flow: The flow in the gap between two concentric, coaxial cylinders of Radii R_1 (inner) and R_2 (outer cylinde) and length L rotating with angular velocities Ω_1, Ω_2.

Thus, important parameters of the TC system are the radius ratio $\eta = R_1/R_2$, the aspect ratio $\Gamma = L/d$ with the gap width $d = R_2 - R_1$ and the ratio of angular velocities $\mu = \Omega_2/\Omega_1$. To characterize the flow in the gap, one can derive two Reynolds numbers related to the inner (1) and outer (2) rotation , $Re_{1,2} = R_{1,2}\Omega_{1,2}d/\nu$, with the kinematic viscosity ν of the fluid. In analogy to other shear flows a shear Reynolds number can be defined as $Re_S = \frac{2}{1+\eta}|Re_1 - \eta Re_2| = \frac{2R_2R_1d}{(R_2+R_1)\nu}|\Omega_2 - \Omega_1|$ [9], which differs from the one introduced in [37] for the boundary layers.

2.2.1 Transport of angular motion

Recently Eckhardt et al. [10, 11] derived a theory on torque scaling in TC flow. This theory describes the transverse current of azimuthal motion J_ω that is constant over all radii with

$$J_\omega = r^3(\langle u_r\omega\rangle_{A,t} - \nu\partial_r \langle\omega\rangle_{A,t}), \tag{2.11}$$

where $\omega = u_\varphi/r$ is the local angular velocity and the brackets $\langle\,\rangle_{A,t}$ denote a mean over a cylindrical surface at radius r and over time. The cylinders are impermeable so the radial velocity and the first term of the equation equal zero at the boundaries $(u_r(r = R_1, R_2) = 0)$. At this radial position the second term is proportional to the torque T that the fluid experiences at the boundary. Hence, measuring the dimensionless torque $G = T/(2\pi L \rho \nu^2)$ at the rotating inner or outer cylinder corresponds to measuring the current $J_\omega = \nu^2 G$. This theory is in close analogy to the Grossmann-Lohse theory on Rayleigh-Bénard flow (RB) [14] and denotes several analogies between the systems [10, 11]. Corresponding to the heat transport Nusselt number in RB one defines a quasi-Nusselt number of TC flow $Nu_\omega = G/G_{lam}$, where $G_{lam} = R_1 R_2 Re_S d^{-2}$ is the dimensionless torque of the analytical Couette solution.

This torque has been measured in the past and has been scaled with the Reynolds number Re_1. Wendt [39] measured the torque at the rotating inner cylinder for three different radius ratios $\eta = 0.68, 0.85, 0.935$ and Reynolds numbers between $50 \leq Re_{1,2} \leq 10^5$. He fitted the dimensionless torque to power laws $G = Re^\alpha$ with $\alpha = 1.5$ for $400 \leq Re_1 \leq 10^4$ and $\alpha = 1.7$ for $10^4 \leq Re_1 \leq 10^5$. Lathrop et al. [20], using a radius ratio of $\eta = 0.724$ for a Reynolds number range of 10^3 to 10^6 and $\mu = 0$, found that the scaling exponent α of the torque still monotonically increases from 1.66 to 1.87 - beyond a critical Reynolds number of 1.3×10^4. Lewis and Swinney [21] repeated these measurements with higher accuracy and verified the results. Later, Paoletti and Lathrop [30] revised the experiment and measured the torque for various flow situation of counter- and co-rotating cylinders. For measurements at a constant shear Reynolds number Re_S they observed a peak in the torque data for a counter-rotation corresponding to $\mu_{max} = -0.33$ for $\eta = 0.725$. Furthermore, the Twente turbulent Taylor-Couette experiment (T^3C) [17, 37, 38] is used to investigate the state of ultimate turbulence in TC flow. Their experiments show that the quasi-Nusselt number Nu_ω scales with the shear Reynolds number like $Nu_\omega \sim Re_S^{0.78 \pm 0.03}$ [17, 38] for different investigated ratios of angular velocities μ. Recompensating this scaling $(Nu_\omega \cdot Re_S^{-0.78})$ they obtain the dependency of Nu_ω on μ. They found a maximum at $\mu_{max} = -0.33 \pm 0.04$ for $\eta = 0.716$ as well. A collaborative work [31] shows how the data of the Maryland and Twente experiment coincide and the torque factorizes into $G = f(\mu) g(Re_S)$ as previously observed in [9, 30, 38]. But the question still remains, how do $f(\mu)$ and $g(Re_S)$ depend on the geometry of the Taylor-Couette systems? A first answer to this question was given by the Twente group by the conjecture that the angle bisector between the two Rayleigh stability lines could serve as a potential prediction for the location of the torque maximum in dependence on the radius ratio. This hypothesis is supported by their measured maximum at $\mu_{max} = -0.33$ and PIV data [17] for $\eta = 0.716$ [17, 37]. The experimental investigation for $\eta = 0.5$ in Merbold et al. [24] are in disagreement with this hypothesis. Brauckmann and Eckhardt [3] as well as the experimental investigations of various other η by Ostilla-Monico et al. [27] show that this prediction is also not valid for narrow gaps. Brauckmann and Eckhardt [3] explained the torque maxima by the enhanced large scale circulation using the analytical expression of the Taylor-Couette stability of Esser and Grossmann [12], for relatively wide gaps this prediction holds with the experimental and numerical findings. We call this prediction for the location of the torque maxima dependent on radius ratio EG-prediction. For the radios ratio $\eta = 0.5$ different velocimetry studies were performed. Van der Veen et al. [36] performed Particle Image Velocimetry in the experiment in Cottbus for different flow behaviours supporting the EG-prediction. These measurements are summarized in Chapter 5. The measurements reported in Chapter 8 and also by Froitzheim et al. [13] were done in the same experiment and extended the parameter space. The experiments, in agreement with the EG-prediction, showed a strengthening of the vortices at the torque maxima while for pure inner cylinder rotation the energy of Taylor-Vortices vanishes for high Reynolds numbers. While the location of the maxima can be determined accurately by the EG-prediction the torque for a given set of rotations is still not predictable.

Chapter 3

The Cottbus Taylor-Couette experiments: Experimental setup and measurement techniques

3.1 Turbulent Taylor Couette Cottbus

To investigate the turbulent Taylor-Couette flow a specially designed experiment is used. A drawing and image of the experiment is shown in Fig. 3.1. The experiment has been used in a prior research project to investigate the influence of a radial temperature gradient onto Taylor-Couette flow. For the present study the experiment needed a crucial revision of the design (chapter 3.1.1). Prior the experiment was able to reach Reynolds numbers up to $Re_S < 10^5$ for inner cylinder rotation. The outer cylinder was kept at rest. To reach very high Reynolds numbers in the order of 10^6, the driving of the inner cylinder had to be changed (see chapter 3.1.3). Also an exchange of bearings and sealings became necessary. The outer cylinder as well as the end plates are now equipped with driving and gearbox. Different transmission ratios are applied to increase the usable range of angular velocities. As one is interested to measure the angular motion flux depending on the input parameters, the implementation of a torque measurement system inside the inner cylinder is a key advantage of this experiment and discussed in chapter 3.1.2.

3.1.1 Geometry and materials

The experimental investigations are carried out in a Taylor-Couette apparatus surrounded by a water bath box as shown in Fig. 3.2. The water bath box (5) and the outer

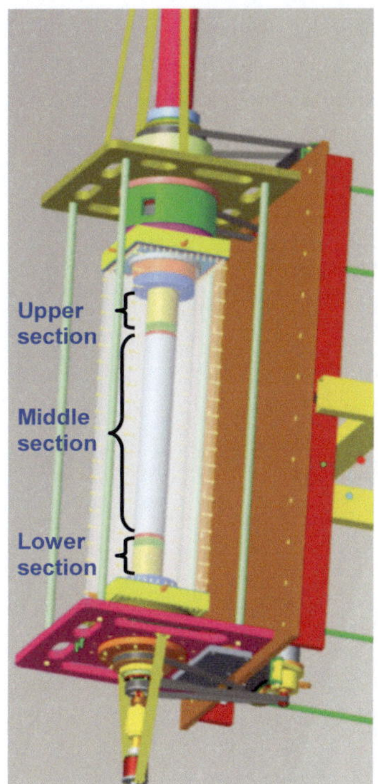

FIGURE 3.1: Drawing and image of the Taylor-Couette system, $R_1 = 35mm, R_2 = 70mm, \eta = 0.5, \Gamma = 20$ with surrounding water bath. The inner cylinder is separated into three sections. The middle section measures the torque. The upper and lower sections are pivoted inside the end plates.

cylinder (2) are made of acrylic glass. The inner cylinder (1) is made of black anodised aluminium. The system radii are $R_1 = 35 \pm 0.02mm$ and $R_2 = 70 \pm 0.2mm$. The end plates (3 and 4) are adjusted to fix the system length to $700 \pm 0.3mm$ by the end plate guidance (10). This leads to a dimensionless radius ratio of $\eta = 0.5$ and an aspect ratio of $\Gamma = 20$. The inner cylinder drives up to a rotation frequencies of $84.5 \pm 0.05Hz$ and the outer cylinder up to $12.2 \pm 0.05Hz$ for the fastest gearing. The inner cylinder has two more gearings to rotate slowly. The outer cylinder does has one more transmission. Both end plates can rotate independently. The influence of the end plate rotation is discussed seperatly. For the majority of measurements of the present study the end plates have been kept stationary. As working fluid we use different silicone oils with kinematic viscosities of $\nu = 0.68 \times 10^{-6}m^2 s^{-1}$, $2.95 \times 10^{-6}m^2 s^{-1}$, $8.41 \times 10^{-6}m^2 s^{-1}$ and $22.81 \times 10^{-6}m^2 s^{-1}$. Therewith we reach shear Reynolds numbers of $10^3 \leq Re_S \leq 10^6$. The temperature of the experiment is controlled to an accuracy of $0.2K$ using the water bath. Velocity measurements by means of Particle Image Velocimetry (PIV) and Laser Doppler Velocimetry (LDV) are only possible through the transparent water bath tank and outer cylinder. Also flow visualisation is been performed in that way.

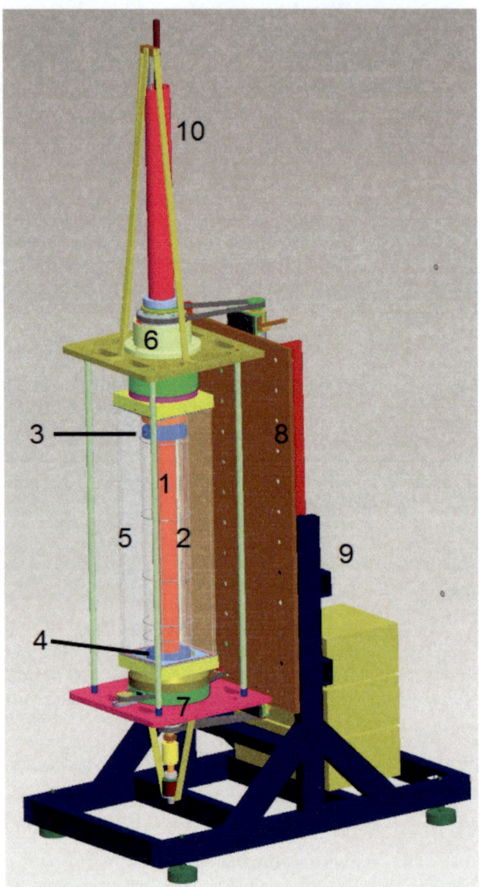

FIGURE 3.2: Detailed sketch of the Taylor-Couette experiment with inner cylinder (1), outer cylinder (2), top end plate (3), bottom end plate (4), water bath (5), top mounting and driving group (6), bottom mounting and driving group (7), mounting plate (8), stativ with weights (9) and top end plate guidance (10).

3.1.2 Torque measurements

We determine the transverse current (Eq. (2.11)) by measuring the torque acting on the inner cylinder according to the flow parameters ($Re_S, \mu, \eta = 0.5, \Gamma = 20$). The inner cylinder is subdivided into three segments (see Fig. 3.3) with lengths $99.4mm$, $500mm$ and $99.4mm$. A gap of $0.6\pm0.05mm$ separates the segments in a similar way as in other TC setups Lathrop et al. [20], Lewis and Swinney [21], Paoletti and Lathrop [30], van Gils et al. [38] (see Fig. 3.1). Only the middle part measures the torque by two strain gauges mounted inside the cylinder. The objective of the separation is to reduce the end effects of the solid end plates onto the torque measurement as in previous experiments [20, 21, 30, 38]. Slip rings transfer the signal to the equalizers (Wachendorf strain gauge converters). The time series (100Hz, 60 seconds) do not show any oscillations or other particular features and fit very well to a Gaussian distribution. The level of standard deviation to mean of the signal is evaluated to be less than 10^{-2}. For the case of larger values the measured torque has been taken as wrongly measured value.

In prior a calibration was necessary. This has been done mounting the inner cylinder without the outer cylinder and water bath. The experiment is adjusted into the horizontal case. A clamp is used mounting two arms at both sides with equal length and weight. At arm length of $500 \pm 1mm$ a grooves is placed. At this groove different weights are hanged. Thus, by gravity a force is acting at arm length of 0.5m and applies a torque to the inner cylinder. Using this method it is possible to caliibrate the torque sensors repeatable to an accuracy of $10^{-4}Nm$ up to a maximal measurable torque of $1.2Nm$. The minimal stable measurable torque is of $2 * 10^{-3}Nm$. If larger torques need to be measured the equalizer amplification can be changed. In this case torques up to $2.4Nm$ are measurable with accuracy of $2 \cdot 10^{-4}Nm$ or up to $4.8Nm$ leading to an accuracy of $4 \cdot 10^{-4}Nm$. The maximal torque observed during experiments using silicone oil of viscosity $\nu = 24cSt$ has been $2.2Nm$. This torque measurement allows us to quantify the transverse current J_ω of equation (2.11) at the inner boundary. With this device we are able to find variations of the radial transport of the flow with the input parameters.

3.1.3   Sytem control

To use the Turbulent Taylor-Couette Cottbus experiment at high rotation rates one needs enough power to drive the machine. The inner cylinder is driven by a 1.3kW motor with maximum rotation number of $2800rpm$. Using a gear transmission this speed is risen to the nominated maximal speed of $5070rpm$. The motor is an asynchron motor driving at a fix frequency wich is adjusted by the power control. The rotation number is kept on a certain speed very stable. In addition to the motor speed also the rotation speed of the inner cylinder is monitored by light barriers. In addition a flexible gearing setting is used consisting of 2 gears for the cylinder driving and two for the motor giving 4 possible transmissions. In that way rotation speeds down to $1 \pm 0.002Hz$ are used. As the outer cylinder has to overcome the drag of the machine, the working fluid as

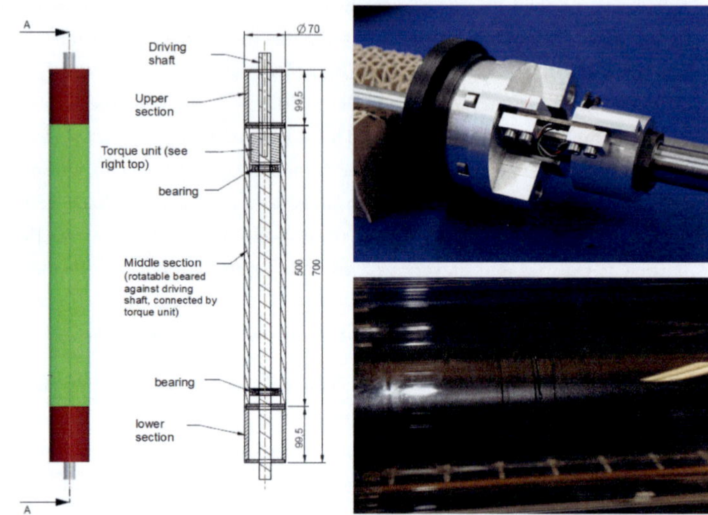

FIGURE 3.3: Drawing and images of the installed torque measurement unit. Left: drawing og the separated segments. The top and bottom red segments as well as the main shaft are driven by the motor. The middle segment (green) is beard against the central shaft and mounted using the strain gauge bridge (c.f. top right). By the use of two symmetrical strain gauges the torque between the rotating shaft and the cylindrical mantle is measured. Top right: Image of the strain gauge bridge and the driving shaft. Bottom right: Image of the final installed cylinder. The gap between the bottom and middle segment is shown.

well as the drag from the water bath a stronger driving is needed for the same rotation speeds. To reach $750 rpm$ maximum speed a 1.5kW motor is used. For high viscous oils this speed can not be reached as the torque becomes to large. For the viscosity of $22.81 \times 10^{-6} m^2 s^{-1}$ the outer cylinder reaches speeds of $410 rpm$. Also this motor is controlled in the same way while the rotation speed of the outer cylinder is monitored by light barriers. To prevent that the outer cylinder gets twisted it is driven parallel from top and bottom mounting. In that way the motor drives a shaft beyound the mounting plate of the Taylor-Couette apparatus which then drives the cylinder. The top and bottom end plates have theire own driving ability. In general each of them could run at unique speed. In the present study both end plates are connected similar to the outer cylinder, so that one motor can drive both at the same speed. The control computer monitors the speed of all rotating parts. By driving the machine the power leads to friction and thus the experiment heats up over time. To prevent this, the surrounding water bath controls the temperature. This is done seperatly by two Lauda cooling baths controlling the temperature of the water bath with an accuracy of $0.05 K$ at temperatures adjusted between $20°C$ and $26°C$. In addition we measure the temperature of the experiment at different locations. One sensor (PT100) is implemented into the rotating inner cylinder and monitors its temperature. Another is placed inside the working fluid below the bottom end plate. These temperatures are used to calculate the fluid properties and to ensure that heat exchange does not affect the experiments.

The control is done using a LabView program. Herewith, the cylinder rotation rates are adjusted. One can choose between a manual change of the rotation speeds or to apply a pre-described linear increase of the rotation rate. The cylinder rotation measured by light barriers, the temperature sensors as well as the strain gauge signals are recorded and monitored while the experiment is running. Using the monitoring of speed and torque it is possible to observe the time that the specific state is established. After several minutes running at one speed there is no change in torque and the fluid flow has been developed. To store the signal one can choose the length of measuring interval in between 1 second and several hours. Long time measurements have been performed for single rotation rates to ensure the stability of the experimental setup. The stability is within a relative standard deviation lesser than 10^{-2}. Later this argument has been used to evaluate torque measurements taken. It is also possible to measure the torque for accelerated cylinder from zero speed to a specific Reynolds number. This drive-up scenario is not in the scope of this work but will be described in a separate study.

3.1.4 Implementation of an artificial wall surface structure

An interesting point for the investigation on torque inside the Taylor-Couette systems is also the question how wall roughness or surface structure does effect the measurements. The system was designed with hydraulicly smooth walls. To see the influence how a wall structure changes the drag at the wall and thus, the measured torques an implementation of a surface structure has been performed. The experimental setup limits the ability of changing the inner cylinder wall. These limits are depicted as:

- The existing inner cylinder geometry has to be used as it is mounted with the strain gauges

- Thus, the inner cylinder radius can not be reduced

- An application of the structure has to be reversible

- The inner cylinder radius should not be changed more than $1mm$ (gap between end plates and inner cylinder)

- The silicone oil and rotation should not influence the changed surface

Different solutions are possible to implement such roughnesses. Two approaches had a technical implementability. One is to use a shrink-on tube the other to usw adhesive tape along the cylinder. As the shrink tube would only enlarge the inner cylinder radius but not give any macroscopic roughness to the system the option of adhesive tape was chosen. By testing different tapes sticked to anodised aluminium and left inside probes of silicone oil it turned out that only OWASAN tape by ThyssenKrupp will be of enough stability for the experiments planned.

The adhesive tape has been attached to the inner cylider in four axial stripes parallel to the axis of the system. Each stripe has a width of $25 \pm 0.5mm$ and was mounted only at the central torque-measuring section with a length of $490 \pm 0.5mm$. The azimuthal separation between each stripe is equally distant and $30mm$. In Figure 3.4 the setup with attached roughness tape is depicted.

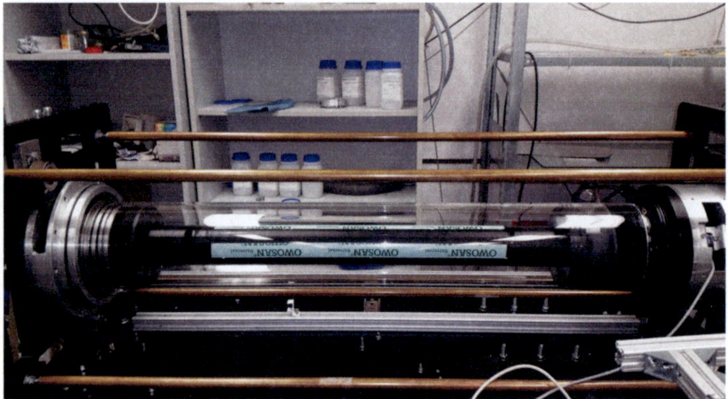

FIGURE 3.4: Inner cylinder of Taylor-Couette system with attached surface structure.

The given adhesive tape does not loosen even for rotation rates up to $3000rpm$ which were used as maximal rotation rate for the given setup. The results of the torque measurements using surface structure is then compared to those with smooth wall.

3.2 Top view Taylor-Couette Cottbus

In addition to the experiment described in section 3.1 we developed a second Taylor Couette apparatus. The main task of the prior experiment is the high accurate measurement of the torque acting on the rotating inner cylinder while reaching shear Reynolds numbers of about 10^6. The change of geometric parameters of the experiment, such as the gap width η, aspect ratio Γ as well as the boundary conditions of the end plates, corresponds to high work load and costs in manufacturing, installation and calibration of the whole experiment. Thus only a few configurations can be made. Also the optical access is limited from the radial direction - so measurements of the radial velocity become unable or a low quality.

Therefore a second Taylor-Couette system is constructed and set up. Its main attributes are: The Variation of the inner cylinder radius and therefore the radius ratio η. The upper and lower end plates can be either mounted to the inner cylinder, outer cylinder or splitted into two rings mounted to each cylinder. Thus the influence of the rotating end plates can be investigated. Beside the outer cylinder also the top end plate is transparent to get optical access for LDV and PIV measurements of the azimuthal and radial velocity components. Torque measurements will be performed using the motor current data. The exchange of the fluid as well as the change of the geometry can be done rapidly.

The geometry of the outer cylinder is exactly the same as for the T2C2 (3.1) to perform comparability of both experiments for the same inner cylinder radius and end plate configurations. A comparison of the dimensionless torque for the same geometry should lead to a calibration of the motor torque measurements by the use of the strain gauge measurements 3.1.2. While also this experiment is rotating on relatively high rotation speeds of both cylinders this causes constructive challenges.

3.2.1 Geometry and materials

A drawing of the Top view Taylor-Couette Cottbus is shown in Fig. 3.5. As described above the radius of the outer cylinder and the length of the system was set to be equal to the one of T2C2 ($R_2 = 70mm, L = 700mm$). The outer cylinder is made out of acrylic glass to get optical access.

The radius of the inner cylinder can be changed. A main shaft of $14mm$ diameter is mounted rotatable inside the outer cylinder. On this shaft different inner cylinders can be mounted. This leads one to the minimal radius ratio of $\eta = 0.1$ while the shaft rotates alone. The upper limit could be in general $R_1 = 69mm$ which would keep a

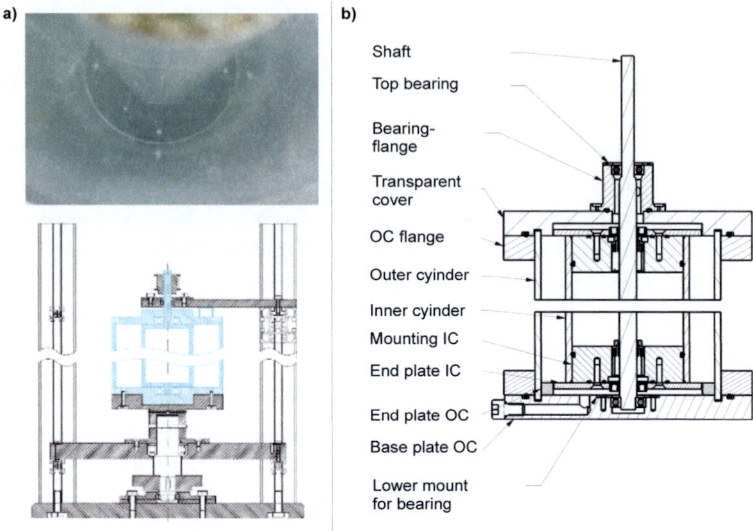

FIGURE 3.5: Top view Taylor-Couette Cottbus. a) View through the top end plate into measurement volume and drawing of the whole setup. A more detailed drawing of the highlighted part is shown in b).

very low accurate gap width due to the manufacturing process of the cylinders. The maximum we built is $R_1 = 50mm$ leading to $\eta = 5/7 \approx 0.71$ which is comparable to many Taylor-Couette systems with torque measurements in literature Brauckmann and Eckhardt [3], Lathrop et al. [20], Lewis and Swinney [21], Paoletti and Lathrop [30], van Gils et al. [37, 38], Wendt [39]. The material of the inner cylinder depends on its radius. For the case of $\eta = 0.1$ the shaft is made from stainless steel. For cases between 0.1 and 0.5 we use different plastics (PET, POM, Polyamid). For the cases larger and equal to 0.5 acrylic glass cylinders are used. In table 3.1 the configurations for this system are presented. Inner and outer cylinder are driven by two independent motors with different gearing possibilities (see section 3.2.2). To measure the torque the fluid acts on the cylinder wall using the torque the motor needs to drive the cylinder it is necessary to decrease all torques made from the machine itself. For this purpose the inner cylinder is beard inside the outer cylinder by only two low friction bearings against the outer cylinder and one against the stativ and has no dynamic shaft sealing. The end plate configuration gives a mounting ability on either the inner as well the outer cylinder. While separating the end plate in two rings of any partition the inner part can rotate with the inner cylinder while the outer part rotates with the outer cylinder. A gap of $0.5mm$ seperates the rings. The partition takes two limits - full end plate rotates with inner cylinder or full endplate rotates with outer cylinder. In the present investigation only the last case is discussed. The influence of the end walls is not being discussed in the previous investigation. The end walls are also made from acrylic glass for the axial optical access, which gives the system its name Top view Taylor-Couette Cottbus (TvTCC).

The main working fluid applied to this experiment is water. Also the use of silicone oils such as in T2C2 is possible as well as running the experiment filled with air. Using air as working fluid the kinematic viscosity becomes higher and therefore we can also investigate low Reynolds numbers with this experiment (i.e. for $\eta = 5/7$ the minimum shear Reynolds number is 80). For this case no torque measurements are possible due to the low friction at the wall while optical measurements can be done. In table 3.1 the geometric variability of this experiment is listed. Different radius ratio η is leading to the listed inner cylinder radi R_1, gap width d, aspect ratio $\Gamma = L/d$, minimum Reynolds number using air (at rotation frequency of n=1/6 Hz) and maximum Reynolds number using water as working fluid (at rotation frequency of n=33Hz). Figure 3.6 shows the $\eta - Re_S$-phase space for our both experiments in comparison to other Taylor-Couette experiments performing torque calculations.

TABLE 3.1: Geometries built in the Top view Taylor-Couette system. The outer cylinder radius is fixed at $R_2 = 70mm$, the length of the system to $L = 700mm$.

η	$d[mm]$	$R_1[mm]$	Γ	$minRe_S$	$maxRe_S$
0.1	63	7	11.11	56	167000
2/7	50	20	14	110	325000
0.5	35	35	20	115	340000
5/7	20	50	35	90	247000

3.2.2 System control

For running the cylinders we use two different motor-gearing configurations. For small rotation frequencies we use two 75 Watt motors of maxon motors leading to cylinder rotation frequencies between 10rpm and 200 rpm. By the use of 250 Watt maxon motors we achieve rotation numbers up to 2600 rpm. The motor controllers are used to set the motor velocity speed. Light barriers are used to observe the cylinders rotation speed. In all cases the velocity speed is constant with an accuracy of less than 0.5%. A LabView program is used to set the motor velocites and calculate the torque the motors need to drive the cylinders.

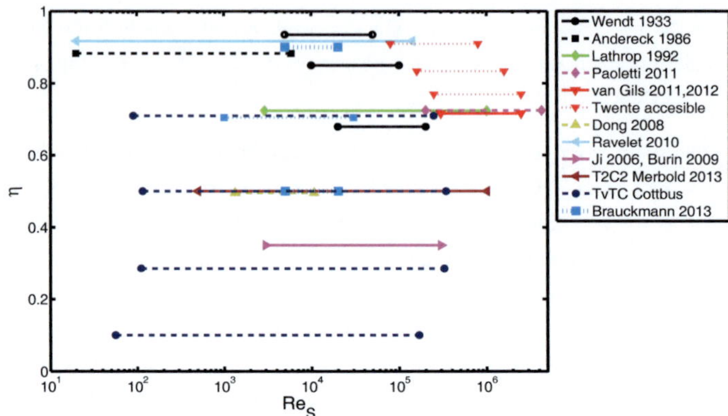

FIGURE 3.6: Phase space of investigations to torque behavior in Taylor-Couette flow for different geometries [1–3, 5, 8, 19–21, 30, 33, 37–39]. Compared to these we show our investigations using the experiment described in section 3.1 as brown solid line and our investigations done with the Top view Taylor-Couete system (dark blue dashed line).

3.2.3 Torque measurement

In addition to the precise torque measurement unit of the T2C2 it was planned to use the motor driving power to evaluate the angular momentum flux also by the TvTCC. The torque the motors need to run the experiment is being measured by its voltage, current and rotation number. By the use of the gearing factor the torque working on the rotating cylinder is being calculated. A calibration of the torque due to the system it self is needed. Measuring the torque of an empty system would give the machines torque produces by mechanical friction. As vacuum is not possible we use the case of an air filled experiment as the drag of the air consists on the wall compared to the one by water is negligible. The torque the fluid effects on the cylinder is the difference to that measured with air. Unfortunately the torque signal examined by the motors power is not stationary and fluctuating hugely. Using the present motors a precise measurement is not possible. Here the experiment still has potential to be designed further to decrease the mechanical drag and increase the ability to measure the torque by the motors directly. Another solution would be to install a shaft-to-shaft torque sensor into the driving pully to be used for torque analisation.

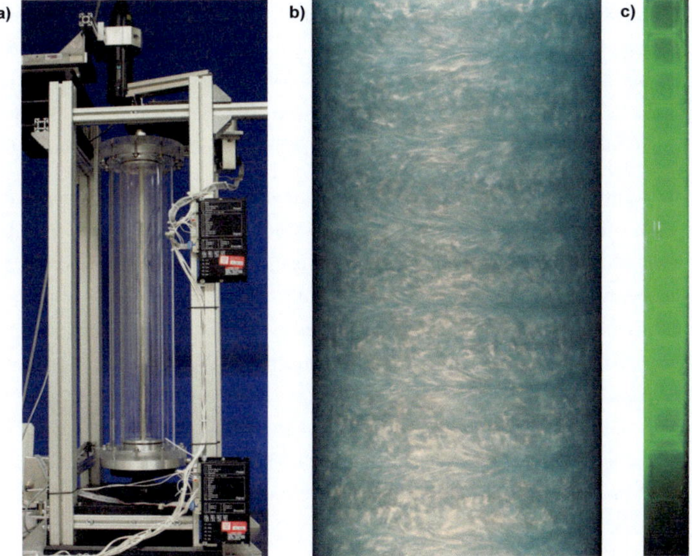

FIGURE 3.7: Top view Taylor-Couette Cottbus. a) Image of the complete setup for $\eta = 5/7$. LDV probe mounted for azimuthal velocity measurements through transparent top plate. b) Visualisation of turbulent Taylor vortices using Kaliroscope particles and blue dye in water for $Re_S = 20,000$ while outer cylinder is at rest ($\mu = 0$). c) Light sheet visualisation of turbulent Taylor vortices at radial-axial plane using smoke in air for $Re_S = 800$ while outer cylinder is at rest.

3.3 Measurement Techniques

In the following chapter the measurement techniques used in this thesis are explained. The optical visualisation methods have been performed to get an image of the ongoing flow features inside the apparatus. Also Particle Image Velocimetry (PIV) and Laser Doppler Velocimetry (LDV) were performed during the measurements. In addition a PIV algorithm was applied to visualisation videos giving a special focus to near-wall velocities.

3.3.1 Access for Optical measurement techniques

As the system is closed and rotating the use of measurement probes such as pressure tubes or hot wire anemometry is, beside the intrusive point of view, an unwanted technique. For that reason the experiment is designed to give optical access for different measurement techniques. As in the most Taylor-Couette systems from Literature the outer cylinder of both experiments gives radial access for visualisation methods. In figure 3.1 the setup of the T2C2 is shown. Due to the bearing and sealing units as well as the outer water bath only the radial optical access is possible. In figure 3.7 the experimental setup of the TvTCC and some visualisation examples are shown. By the use of visualisation methods one can observe the structures inside the system (Fig. 3.7 b). Observations for this case are made systematically and discussed in Chapter 6.

In addition the TvTCC also has a transparent top end plate which makes it unique. This makes it possible to use a Laser Doppler Velocimetry (LDV) probe from top to measure the azimuthal and radial velocity components. The axial component can be measured through the transparent cylinder wall additionally by a second LDV probe. By applying a Laser sheet in azimuthal-radial plane one is able to capture particle images through the top plate. By this Particle Image Velocimetry (PIV) of the radial-azimuthal velocities is possible without refraction problems at the cylindrical surface. In Fig. 3.7 c) the gap is illuminated by an axial-radial light sheet. Oil droplets of $1 - 3\mu m$ size, which are suitable for LDV measurements and light sheet visualisation are used. The LDV and PIV measurements of radial velocity profiles are discusses in sections 3.3.3 and 3.3.4.

3.3.2 Visualisation methods

To understand the global flow structures in turbulent Taylor-Couette systems the use of visualisation techniques is a key issue. It leads to the possibility to observe the Taylor vortices and other flow behaviour. One becomes able to determine the number of vortices, thus to measure theire axial length in terms of gap widths. Also it becomes able to distinguish if the flow is turbulent or not. To do so one needs to add particles which make the flow structures observable. In the present experiments Kalliroscope is used with a typical dimension of approximately $30 \times 6 \times 0.07\mu m$ guanin flakes. This makes them as small to follow the fluid with its velocity. As they are flat they collimate with the local shear stress and reflect light depending on their orientation. If the wide side is directed to the observer the most light is reflected, areas where the particles align in that way, that the sharp edge points to the observer, results in a darker zone. The name represents three greek words "kalos", "rheos" and "skopien" meaning "beauty", "flow" and "seeing". Only a small amount of particles is needed, in our experiments the volume fraction of Kalliroscope compared to the working fluid is kept less than 10^{-4}.

Observing the TC-flow through the outer cylinder, this means an image will be bright where the velocity is axial-azimuthal dominated, while dark areas represent a strong radial-azimuthal flow. This can be seen in figure 3.7 b. In the experiments a blue dye is added to the fluid to increase the contrast of the visualisation images. By the concentration of this blue dye it can be possible to make inner regions of the Taylor-Couette gap invisible. In that way only a wall near region close to the outer cylinder is being observed. This becomes important for counter rotating cylinders in Chapters 6 and 7.

3.3.3 Laser Doppler Velocimetry

A well known method to investigate flow velocities in one position is the Laser Doppler Velocimetry. For this method a monochromatic Laser beam is split into two beams, the two beams are directed into the position the measurement should take place, where they cross each other under a certain angle. In the crossing volume an interference takes place. If a particle is now inside this volume it reflects the interference, the reflection signal can be detected using a photo multiplier. If the particle is in motion along the interference pattern a time dependent signal is observable, the doppler effect shifts the frequency. This shift is dependent on the used wave length of the Laser, the angle of crossing of the two beams and the velocity of the crossing particle. As the first both are kept constant and known one is able to measure the particle velocity by analysing the Doppler frequency. A Bragg cell usually is also used resulting in a movng interference pattern, allowing also the detection in which direction the particle is moving. If the particle is ideally following the flow, it is possible to measure the flow velocity by each event a particle crosses the interference volume. Nowadays this method is widely spread and developed and it can be applied quiet simple by the use of commercial systems. In this thesis a LDV system of Dantec Dynamics has been used to measure the azimuthal velocity component inside the Taylor-Couette gap. In first case the Laser is poined through the acrylic glass cylinder of the TTCC experiment and traversed in radial direction at the mid height of the system. As the angle of the beams is changed due to the refraction inside the system the measured velocities have to be corrected. To detect the correct angle the system was traversed along the radius between inner and outer cylinders, with a gap width of $35mm$. Due to the refraction the path outisde the system was measured by a micrometer screw, resulting in a shift of $28.4mm$. By simple triangulation the effective angle inside th system could be calculated and considered for the velocity measurements. In the second case the Laser is pointed through the transparent top plate of the TvTCC experiment and shiftet by a traverse system along the radius and height, shown in Figure 3.8. Azimuthal velocity as well as radial velocity can be masured in this setup. By the use of the radial velocity at mid gap along the height the position of the turbulent Taylor vortices can be detected. Measurements of radial profiles then were applied at the inflow, the outflow and central heights along one vortex.

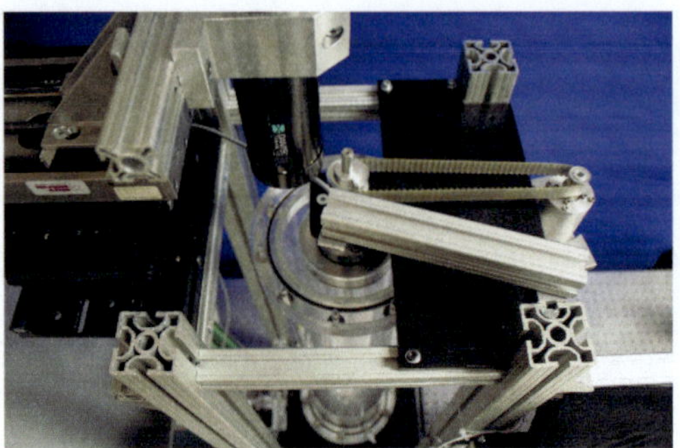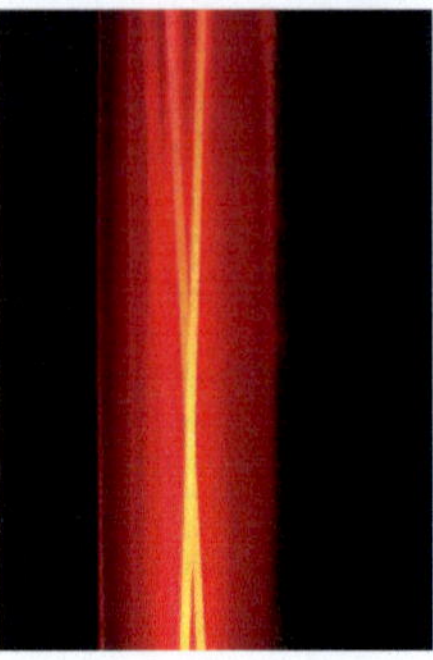

FIGURE 3.8: Laser Doppler Velocimetry used at Top view Taylor-Couette apparatus. Left: LDV probe mounted on top of the experiment at a two dimensional micrometer traverse pointing into the Taylor-Couette gap through the transparent top plate. Right: Image of the crossing area of the both beams inside the gap. Here the case of $\eta = 5/7$ is used.

3.3.4 Particle Image Velocimetry

Another well established measurement method in fluid mechanics is the Particle Image Velocimetry. This method is developed to measure a full instantaneous flow field. Also in this case particles are needed in the flow which we consider to follow the flow perfectly. Thus they need to have a density close to the one of the fluid to reduce buyency effects and need to be as small, that the fluid can transport them without slip. A camera is used to make two photographs of the particles after each other with a time difference of Δt. To achive partice images in a certain measurement plane a huge intensity of light illumination is requiered that the particles scatter light detectable by the camera. For this purpose a Laser pulse is required for each capturing of an image. The Laser beam usually is formed into a light sheet by the use of cylindrical lense optics, as done also in our setups. The time between the two Laser pulses finally defines the time of the two captured particle positions. The displacement, due to the movement of the particle, can be seen between the two images if the time between the pulses is optimal. The time Δt should be arranged, that the displacement represents 4 to $10 pixels$. The optimal camera view is perpendicular to the laser sheet. If the view angle deviates from the perpendicular view the focussing to the particles in the light sheet is less effective but more problematic is, that the measurement will be strongly disturbed by the velocity component out of plane. In stereoscopic PIV a second camera is added and this effect is used to analyse all three velocity components in the plane. In the measurements taken in this thesis the optical access was to limited to perform a stereoscopic PIV. By splitting the image into interrogation areas (IA) the velocity of a patch of particles can

be calculated by using cross correlation between the two IA in time. For each IA in space
a two dimensional velocity vector results. The size of the IA can be changed resulting
in different spatial resolution and accuracies. The method is widely investigated and
a fundamental background is given in [32]. In our case the Top view Taylor-Couette
Cottbus experiment is used. The Laser sheet is pointed in radial-azimuthal plane, the
axial position can be shifted by a traverse system. Through the transparent top plate
the camera is observing the particles in the light sheet by an optimal angle of 90 deg to
the light sheet. Using this setup about 1/3 of the azimutal length can be observed by
means of PIV. Another setup of PIV is also used in this thesis. In Taylor-Couette it
is of interest to understand the azimuthal-axial flow at a cylindrical surface (compare
to theory in Chapter 2). The setup used reflecting particles, a strong light source from
outside and a high concentration of dye in the fluid. The cameras view was through the
transparent outer cylinder. Thus, only a thin layer of fluid could be seen close to the
outer cylinder. The dye stopped the view into the system within $2mm$. As the flow close
to the outer cylinder is relatively slow, this continous light combined with the camera
frame-rate delivered good particle images and displacements to be analysed by PIV up
to $Re_S = 7.8 \cdot 10^4$ while it worked optimal for $Re_S = 2.5 \cdot 10^4$.

Chapter 4

Experimental determination and empirical prediction of torque

4.1 Torque measurements

As described before one of the main abilities of our experiment is the precise torque measurement inside the inner cylinder. For various flow states the torque was measured and will be discussed in this section. The first experiments were done with outer cylinder at rest. For this case the inner cylinder rotation rate was increased and decreased for a wide range of shear Reynolds number. Especially experiments for counter rotating cylinders have been explored. For these cases the shear Reynolds number was kept constant by fixing $\Delta\Omega = \Omega_2 - \Omega_1$, while the ratio of angular velocities $\mu = \Omega_2/\Omega_1$ was varied. This has been done for several constant values of the shear Reynolds number. The investigated phase space in terms of shear Reynolds number and angular velocities is given in Fig. 4.1.

The experimental procedure followed allways the same protocol. First the inner cylinder was accelerated to a given shear Reynolds number. As the number of vortices does depend on many parameters of a system it was taken care to get 20 vortices for the start of these experiments. If a different number of vortices was observed, the system was slowed down and accelerated again. Torque measurements for this case (Given Re_S while μ equals 0) were done. Afterwards the inner cylinder has been slowed down while the outer cylinder was switched to counter rotation leading to the same $\Delta\Omega$. After giving the fluid enough time to establish the flow, the torque was measured for this case. This procedure has been done up to the limit of cylinder rotation rates.

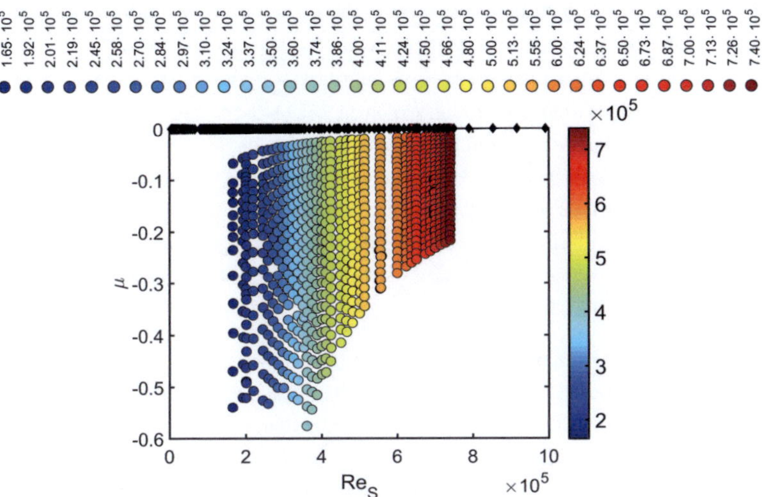

FIGURE 4.1: Phase space of the experimentally measured torques discussed in the following. For given shear Reynolds numbers Re_S between $1.65 \cdot 10^5$ and $7.6 \cdot 10^5$ the ratio of angular velocities μ has been varied. Also for stationary outer cylinder $\mu = 0$ the shear Reynolds number was varied with higher resolution

Investigations of the case of pure inner rotation as well as different cases of co- and counter-rotation have been performed (Fig. 4.1). The measured quasi-Nusselt numbers $\mathrm{Nu}_\omega = G/G_{lam}$ are shown in Fig. 4.2 for different cases of flow configurations. The classical case where the outer cylinder is at rest ($\mu = 0$) is indicated in black. The region of co-rotation is distinguished by the Rayleigh stability criterion $\mu = \eta^2$, while the counter-rotation is additionally distinguished by $\mu = -0.2$, where the observed maximum in torque is approximately located for these experiments. The scaling of the dimensionless torque with shear Reynolds number $G \sim Re_S^\alpha$ is calculated using the $\mu = 0$ data. Here, one could observe two regimes with scaling exponent of $\alpha = 1.62 \pm 0.04$ for shear Reynolds numbers $3 \times 10^3 \leq Re_S \leq 8 \times 10^4$. For Reynolds numbers $10^5 \leq Re_S \leq 10^6$ we find a scaling exponent of $\alpha = 1.78 \pm 0.3$ similar to the one in the Twente experiment ($\alpha = 1.78 \pm 0.06$). Hence, the Re_S dependence of the exponent becomes important. The dependence is also shown in Fig. 4.2, calculated using a sliding least square fit over different ranges of $\Delta(log_{10}Re_S) = \Delta_{10}$. For lower shear Reynolds numbers ($Re_S \leq 8 \times 10^4$) the fluctuations of the exponent are strong which is reflected in the correlation coefficient of $R = 0.97$ for $\Delta_{10} = 0.2$. For a larger regression interval this becomes less important and R exceeds 0.99. At high shear Reynolds numbers ($Re_S > 10^5$) the local exponent calculated using the largest interval differs from the other two local exponents which coincide. Here, the correlation coefficient for $\Delta_{10} = 1$

decreases. The calculated exponent in this regime does not appear to be correct anymore, while for $\Delta_{10} = 0.2, 0.6$ the correlation coefficient becomes $R > 0.995$. Hence, the calculation using different intervals influences the interpretation of the scaling exponent. We conclude that the scaling exponent calculated using $\Delta_{10} = 0.6$ is closest to the real behavior over the full range of shear Reynolds numbers. The scaling exponent for shear Reynolds numbers $\mathrm{Re}_S > 6 \times 10^5$ seems to tend to a constant value of about $\alpha = 1.75 \pm 0.03$ for $\eta = 0.5$ which compares well with the exponent of 1.78 ± 0.06 by van Gils et al. [37]. Thus keeping the relatively high uncertainty of this exponent in mind, the scaling exponent seems not to depend on the curvature of the Taylor-Couette system for high Re_S.

In the lower Reynolds number regime ($10^3 \leq \mathrm{Re}_S \leq 2 \times 10^4$) we are able to directly compare the measured torques with calculations from direct numerical simulations with the same η but different $\Gamma = 2$ carried out by Hannes Brauckmann [2]. These simulations are using a Fourier-Chebyshev spectral code [25] with periodic boundary conditions in axial direction. These computations reach up to $\mathrm{Re}_S = 2 \times 10^4$ and satisfy three convergence criteria as explained in a study for $\eta = 0.71$ by Brauckmann and Eckhardt [2]. For $\mathrm{Re}_S = 10^4$ and $\mathrm{Re}_S = 2 \times 10^4$ only one third of the domain in azimuthal direction is simulated with periodic boundary conditions. We tested that the azimuthally restricted domain does not bias the torque computation. To reach these relatively small values with our experimental setup it is necessary to use the highly viscous silicone oil to still measure an accessible torque amplitude.

We compare torques for three shear Reynolds numbers $\mathrm{Re}_S/10^3 = 5.0, 10, 20$ and different ratios of angular velocities μ. In Fig. 4.3 the numerical investigation is indicated by triangles and compared to measured torques for two different fluids (circles, crosses). The numerical and experimental studies show a good agreement with each other in the Rayleigh-unstable regime. The relative deviation does not exceed 4% for counter-rotating cylinders. Consequently, the restricted numerical domain ($\Gamma = 2$ and only one third in azimuthal direction) carries already the essential turbulent characteristic to feature the same transport behavior. The deviations between the numerical and experimental torques at $\mathrm{Re}_S = 1.00 \times 10^4$ for counter-rotation $\mu < -0.18$ are likely due to different wavelengths of Taylor vortices. The numerical domain selects a wavelength of $2d$, while the experiment was not able to reproduce this wavelength for these rotation rates. The wavelength became $2.22d$ for $\mu < -0.18$ while for angular velocity ratios of $\mu > -0.18$ the experiments give a wavelength of $2d$. The influence of different wavelengths has been discussed already by Lathrop et al., Lewis and Swinney [20, 21] and recently in a numerical study [2].

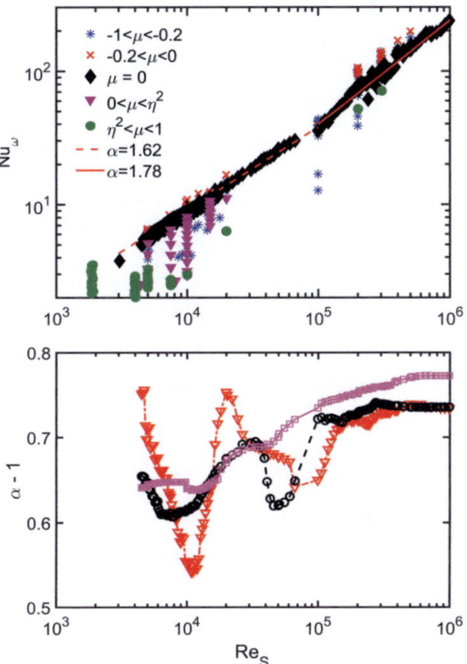

FIGURE 4.2: a) The dependency of the quasi-Nusselt number $Nu_\omega = G/G_{lam}$ on the shear Reynolds number Re_S over the whole range of measurements. Strong counter-rotation ($\mu < -0.2$) is indicated by stars, counter-rotation of $-0.2 < \mu < 0$ by crosses, inner cylinder-rotation $\mu = 0$ is indicated by dots, co-rotation at Rayleigh-instable regime $0 < \mu < 0.25$ triangles and co-rotation at Rayleigh stable regime $\mu > 0.25$ by circles. The dashed line indicates a power law fit $Nu_\omega^{\alpha-1}$ with the exponent $\alpha = 1.62$ for $3 \times 10^3 \leq Re_S \leq 8 \times 10^4$ and $\alpha = 1.78$ for $10^5 \leq Re_S \leq 10^6$ (solid line). b) Local exponent for the quasi-Nusselt number $\alpha - 1 = \partial(log_{10}Nu_\omega)/\partial(log_{10}Re_S)$ in dependence on the shear Reynolds number, calculated from torque measurements for $\mu = 0$ using a sliding least square fit over intervals of $\Delta(log_{10}Re_S) = \Delta_{10} = 0.2$ (triangles), 0.4 (circles), 1 (squares).

Obviously, the torque in the experimental and numerical study is maximal at a rotation ratio of $\mu_{max} = -0.2$. Furthermore, the torque measurements for a selection of shear Reynolds numbers $10^5 \leq Re_S \leq 5 \times 10^5$, where no numerical calculations were possible, are shown in Fig 4.4. The full set of data up to shear Reynolds number of $Re_S = 7.4 \times 10^5$ is given in Fig. 4.5. The torque for this various number of measurements show that the maximum remains located at $\mu_{max} = -0.20 \pm 0.02$ for all shear Reynolds numbers in this investigation. In contrast van Gils et al. [37] suggested, using the bisection of the unstable parameter regime, that the maximum is located at $\mu_{bis} = -0.31$ for $\eta = 0.5$.

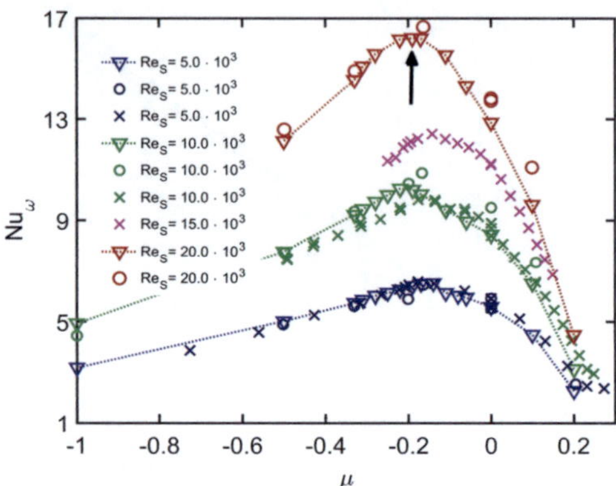

FIGURE 4.3: The dependency of the quasi-Nusselt number $Nu_\omega = G/G_{lam}$ on the ratio of rotation rates μ for constant shear Reynolds numbers ($Re_S/10^4 = 0.5, 1, 1.5, 2$). The circles represent experimental torque data for silicone oil $\nu = 8.41 \times 10^{-6} m^2/s$, the crosses represent measurements for $\nu = 22.81 \times 10^{-6} m^2/s$ compared with direct numerical simulations (triangles). Dotted lines serve as guide to the eye, the black arrow indicates the maximum at $\mu = -0.2$.

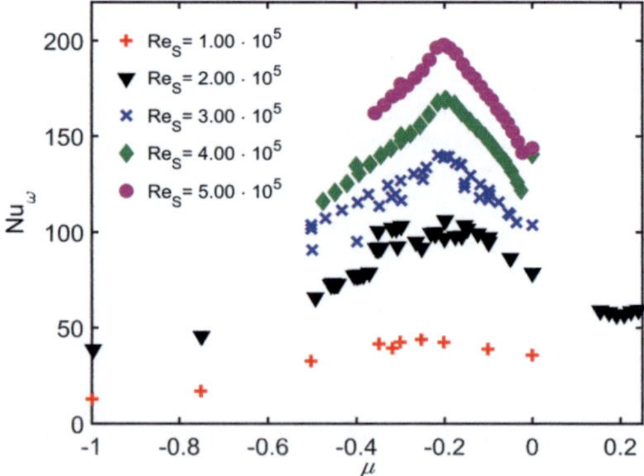

FIGURE 4.4: The dependency of the quasi-Nusselt number $Nu_\omega = G/G_{lam}$ on the ratio of rotation rates μ for constant shear Reynolds numbers ($Re_S/10^5 = 1.00, 2.00, 3.00, 4.00, 5.00$).

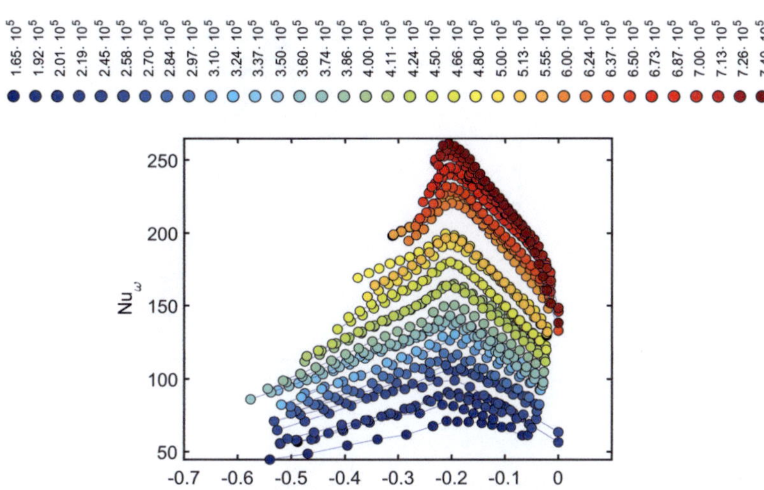

FIGURE 4.5: The dependency of the quasi-Nusselt number $\mathrm{Nu}_\omega = G/G_{lam}$ on the ratio of rotation rates μ for constant shear Reynolds numbers ($\mathrm{Re}_S/10^5 = 1.65$ up to 7.40)

Thus the angle bisector does not coincide with the torque maximum for TC with radius ratio $\eta = 0.5$.

For Reynolds numbers $\mathrm{Re}_S/10^5 = 2.00$ and 3.00 one observes two branches for the torque in the range of $-1 \leq \mu \leq -0.1$, cf. Fig. 4.4. These two branches are not failures during the measurement procedure since they are reproducible. They result from the behavior of the flow. This suggests that the flow structures change between different turbulent flow states existing in this parameter regime. When the cylinders are rotating faster than $10Hz$, as they do here, it is nontrivial to observe the two different flows. A discussion of the flow behaviour depending on the ratio of angular velocities follows in section 6.1.

Compensating all torque data for $Re_S > 1.6 \times 10^5$ by dividing the Nusselt numbers by the Nusselt number of the maximum for each shear Reynolds number depending on μ the curves intersect roughly (Fig. 4.6). If all the compensated torques interesect it is possible to claim, that the Nusselt number behaviour with shear Reynolds number and ratio of angular velocities $Nu_\omega(Re_S, \mu)$ can be split into two functions $f(ReS)$ and $g(\mu)$. The function $f(ReS)$ can be found by the behaviour of the Nusselt number against shear Reynolds number for outer cylinder at rest discussed already [24]. The second part $g(\mu)$ implies a more complex dependency. The compensated torques o show

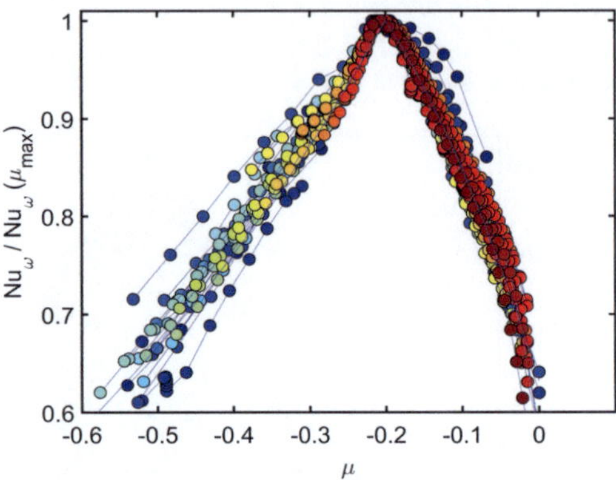

FIGURE 4.6: The dependency of the quasi-Nusselt number $Nu_\omega = G/G_{lam}$ on the ratio of rotation rates μ for constant shear Reynolds numbers ($Re_S/10^5 = 1.65$ up to 7.40) compensated by the maximum of each shear Reynolds number.

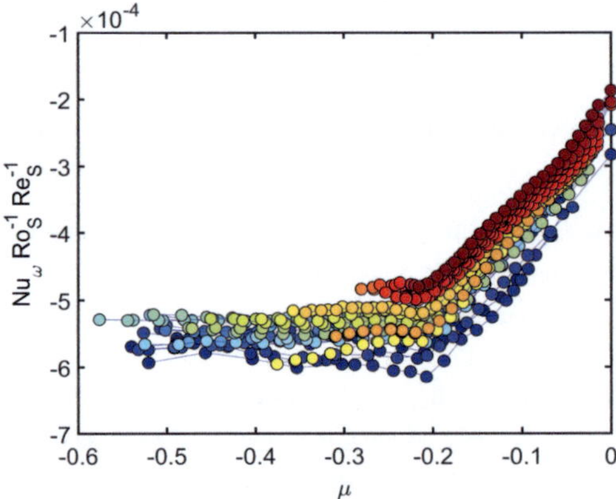

FIGURE 4.7: Quasi Nusselt number devided by shear Reynolds number and the Rotation number $NuRe_S^{-1}Ro_s^{-1}$ in dependence on μ.

a general behaviour but the experimental deviations need to be considered. In the following chapter the behaviour of the torque is discussed in a different perspective.

4.2   Torque prediction

Using the various measurements of the quasi Nusselt numbers it is possbile to detect the dependency of the torque with the input parameters. This is of huge impact for

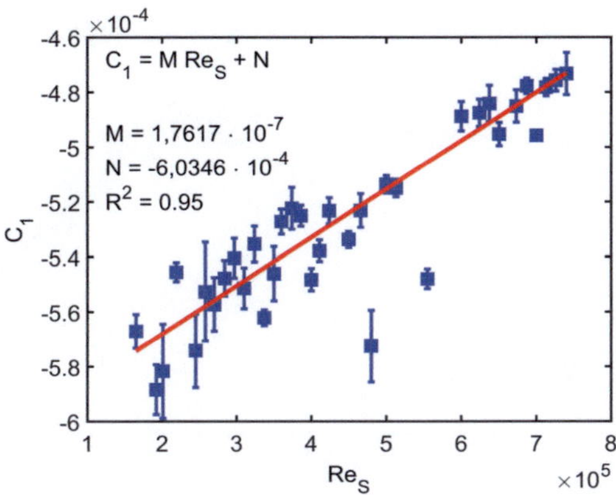

FIGURE 4.8: Linear regression of C_1 the compensated Torque $Nu_\omega Re_S^{-1} Ro_S^{-1}$ averaged for interval $-1 <= \mu <= \mu_{max}$ depending on the shear Reynolds number Re_S.

technical applications, as the torque determines the driving power for rotating devices. It has already been shown that the torque strongly increases with the shear Reynolds number and has a strong dependency on the ratio of the angular velocities. To reveal the dependency of the torque, an additional rotation parameter is used, defined as: $Ro_S = \frac{d}{\Delta\omega}(\frac{\omega_2}{R_2} + \frac{\omega_1}{R_1})$, which represents the mean of the angular velocities in relation to the radii. In Fig. 4.7 the quasi Nusselt number is compensated by the shear Reynolds number Re_s (which is linear to the laminar torque, $G_{lam} = R_1 R_2 d^{-2} Re_S$) as well as the mean rotation number Ro_S and given in dependency on the ratio of angular velocities. Obviously the Nusselt number strongly depends on these both parameters: $Nu_\omega \sim Re_S \cdot Ro_S$. The dependency on the ratio of angular velocities μ for rotation ratios beyond the torque maximum ($\mu < \mu_{max} = -0.2$) disappears and follows a linear relation for rotation ratios in between torque maximum and pure inner cylinder rotation ($\mu_{max} < \mu < 0$). A weak dependency on the shear Reynolds number still can be observed by the scattering of the data. To analyse this dependency for strong counter rotation the values of $Nu_\omega Re_S^{-1} Ro_S^{-1}$ are averaged in between $-1 <= \mu <= \mu_{max}$, defining the coefficient C_1. The mean and its standard deviation are given as coefficient C_1 in Fig. 4.8 in dependence on the shear Reynolds number. Obviously one can observe the linear increase of the coefficient with the shear Reynolds number for strong counter rotation:

$$Nu_\omega Re_S^{-1} Ro_S^{-1} = C_1 = M Re_S + N \tag{4.1}$$

A linear fit with the square of correlation coefficient $R^2 = 0.95$ can be used to estimate this function, leading to the predictable torque with slope $M = 8.8 \cdot 10^{-11}$? and offset

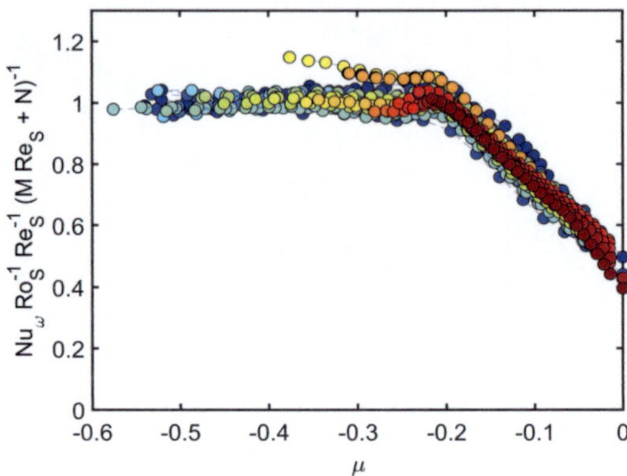

FIGURE 4.9: The Torque data compensated by equation 4.2 for strong counter rotation $\mu < -0.2$. The measured data intersects mainly within an interval of $5 \cdot 10^{-2}$ in relation to the predicted torque, a few outlier deviate up to max 10^{-1}.

$N = -3.3 \cdot 10^{-4}$. Concluding this, the dimensionless torque $G = Nu_\omega G_{lam}$ for strong counter rotation can be given by the formula:

$$G = R_1 R_2 d^{-2} (M Re_S + N) Re_S^2 Ro_S \tag{4.2}$$

$$T = 2\pi L R_1 R_2 d^{-2} \rho \nu^2 (M Re_S + N) Re_S^2 Ro_S \tag{4.3}$$

The quotient of the measured torque and the predicted torque is given in Fig. 4.9 and shows the quality of the torque prediction for the strong counter rotation

As described above the behaviour of the torque for slight counter rotation also depends on a linear increase with the ratio of angular velocities μ, see Fig, 4.10. This secondary linear behaviour is analysed by linear regression independent of the shear Reynolds number. Thus, the torque for $-0.2 < \mu < 0$ can be described by:

$$G = R_1 R_2 d^{-2} (M Re_S + N) Re_S^2 Ro_S \cdot (O\mu + P) \tag{4.4}$$

$$T = 2\pi L R_1 R_2 d^{-2} \rho \nu^2 (M Re_S + N) Re_S^2 Ro_S \cdot (O\mu + P) \tag{4.5}$$

A linear regression of the data of Fig. 4.10 results in the regression coeffients: $O = -2.6367, P = 0.47401$.

The quotient of the measured and predicted torque (Eq. 4.4 and 4.5) for weak counter rotation $-0.2 < \mu < 0$ is calculated to examine the deviations of the prediction. These relative deviations are given in Fig. 4.11. It is shown that the torque can be predicted

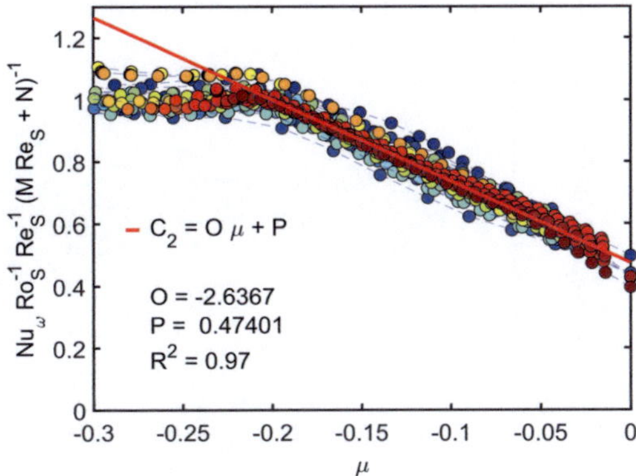

FIGURE 4.10: The Torque data compensated by equation 4.2 for weak counter rotation $-0.3 < \mu < 0$. Linear regression of torque data compensated by equation 4.2 for weak counter rotation $-0.2 < \mu < 0$. $M_2 = -2.6367$, $N_2 = 0.47401$

by this method with an accuracy of 10%. Thus, the prediction made here is relevant for technical applications where torque plays a crucial role in rapidly rotating machines.

4.2.1 Conclusion

The torque acting on rapidly rotating cylinders in a Taylor-Couette flow for radius ratio $\eta = 0.5$ is measured for counter rotating cylinders in the range of the shear Reynolds number $1.65 \cdot 10^5 < Re_S < 7.5 \cdot 10^5$. We observed a strong dependence of the torque with the mean of relative angular velocities and shear Reynolds number: $G \sim Re_S^2 Ro_S$. Using this we found an empirical equation for the prediction of the torque leading to a precise estimation of the torque in a Taylor-Couette geometry with radius ratio of $\eta = 0.5$. This prediction holds within an accuracy of 5% for flow regime beyond the torque maximum $\mu < \mu_{max}$ and deviationd of less than 10% for flow regimes of weak counter rotation $\mu_{max} < \mu < 0$.

4.3 Torque for Co-Rotating Flow

Apart from the main objective of this investigation also some torque measurements were taken in the co-rotating and centrifugally stable regime. As already discussed, the torque observed did decrease for increasing rotation ratio $\mu > -0.20$ for this experiment. In Figure 4.12 the behaviour in the co-rotating regime is depicted. By the knowledge of the centrifugal instability one would expect the quasi Nusselt number to drop closely to 1

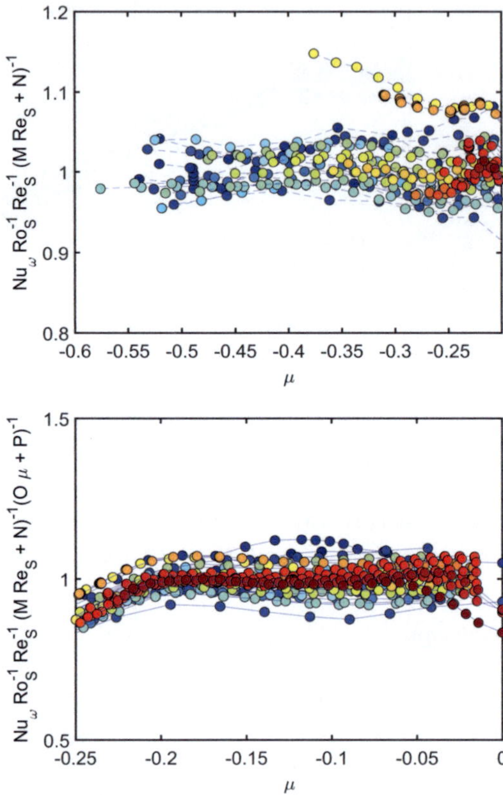

FIGURE 4.11: Top: The Torque data compensated by equation 4.2 for strong counter rotation $\mu < -0.2$. The measured data intersects mainly within an interval of $5 \cdot 10^{-2}$ in relation to the predicted torque, a few outlier deviate up to max 10^{-1}. Bottom: Torque data compensated by equation 4.4 for weak counter rotation $-0.3 < \mu < 0$.

when the Rayleigh stability line is passed. Interestingly it does increase for the following rotation rates. The torque has its minimum in the centrifugally instable regime. Of course for rotation ratio $\mu = +1$ the system goes into the case of solid body rotation. In the case of solid body rotation the system has no shear and thus, no drag. There the torque is minimal again. Thus, there must be a second maximum. And this maximum must be located in the quasi-Keplerian regime, which is between the Rayleigh-stability line and the solid body rotation.

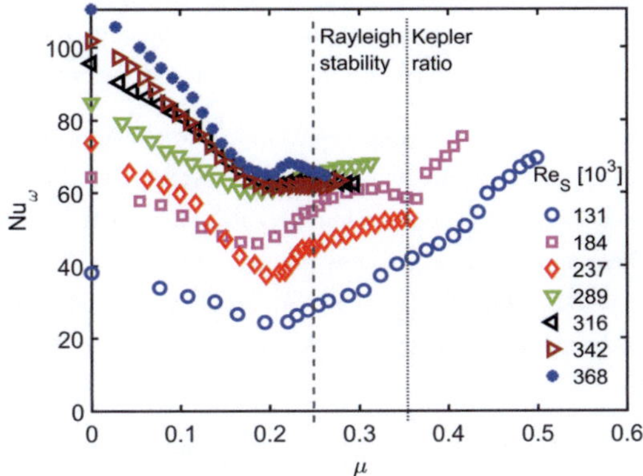

FIGURE 4.12: The dependency of the quasi-Nusselt number $Nu_\omega = G/G_{lam}$ on the ratio of rotation rates μ for constant shear Reynolds numbers ($Re_S/10^5 = 1.3$ up to 3.7) for co-rotating cylinders. The dashed lines indicate the Rayleigh stability criterion $\mu = \eta^2$ and the Kepler rotation ratio $\mu = \eta^{3/2}$.

One explanation of these high torques would be effects driven by the end plates growing into the entire gap. But the torques measured are huge for such known flow structures. One can expect, that for the turbulent case the flow behaviour is again full of secrets one can investigate. Unfortunately the scope of this thesis doesn't allow further measurements to study the torques and its reason in co-rotating system. Thus, further investigations can reveal this interesting field of study.

4.4   Surface structure

The experiment also has the ability to study how the form of the wall surface influences the torque.

Smooth M	=	The torque data published in [24] and already presented	
2013			
Smooth ν_1	=	Smooth surface measurement 1	
Smooth ν_2	=	Smooth surface measurement 2	
Rough ν_1	=	Surface structure measurement 1	
Rough ν_2	=	Surface structure measurement 2	
Rough ν_3	=	Surface structure measurement 3	
Rough ν_4	=	Surface structure measurement 4	

The different measurements are done using different silicone oils with slightly different viscosity. Figure 4.13 the results for these cases are shown. In the Reynolds range of

$3 \times 10^3 \leq Re_s \leq 4 \times 10^4$ the differences of the dimensionless torques of smooth and structured wall are small. Increasing the Reynolds number $4 \times 10^4 \leq Re_s \leq 8 \times 10^4$ the slope for the structured surface becomes stronger and the torques in both cases differ.

To increase the maximal Reynolds numbers in these measurements the system gearing had to be changed and cylinders were turned faster ($1.3 \times 10^4 \leq Re_s \leq 1.2 \times 10^5$). It has to be mentioned that after the highest rotation rates the adhesive tape experienced a to high shear and it started to loose from the cylinder surface. This was observed and the measurements stopped as soon as this happened. In the case of structured surface the slope of torque in the range of $9 \times 10^4 \leq Re_s \leq 1.2 \times 10^5$ becomes immense, see Figure 4.13.

Now the system transports angular motion much more effective. Figuring the quasi-Nusselt numbers 4.13 the transport becomes three times as strong as for the smooth case. For lower Reynolds numbers the torque of this curve is smaller compared to the smooth case. As no flow further flow measurements were able during these measurements a clear explanation for this can not be given. But in agreement with measurements of the smooth cases using different gearing and acceleration ramps different numbers of Taylor-Vortices were observed leading to different torque signals. Such behaviour is also possible in the present case as the acceleration to the smallest measured Reynolds number for this case was much faster due to the motor drive-up.

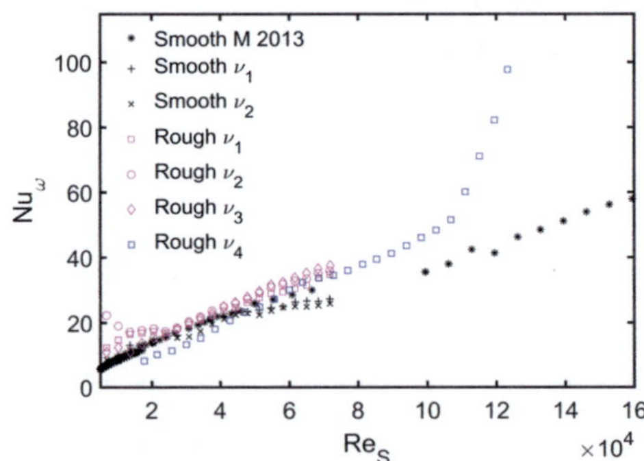

FIGURE 4.13: Comparison of the quasi-Nusselt number against shear Reynolds number for smooth and structured surface for different measurements

The influence of structured surface has been given with these measurements. A more detailed study can be performed and is also planned. As it was not the scope of the present thesis no further geometries of wall structures are studied.

Chapter 5

Velocity measurements

5.1 Angular velocity profiles at mid hight

Inside the Top view Taylor-Couette system with transparent end plate, we applied a PIV measurement. The light sheet illuminates an azimuthal-radial plane in the mid-hight of the experiment, we observe the plane through the upper end. The field of view is about more than one fifth of the whole gap (see Fig. 5.1). The measured two dimensional, time averaged cartesian flow field is interpolated into a cylindrical coordinate system. The shown profiles of the angular velocities ω (Fig. 5.1, 5.2) are averaged over $2\pi/5$ for different radii and represented in comparison to the analytic laminar Couette profile ($\omega(r) = Ar + B/r$, where A and B are geometric factors). The flow in all these cases is turbulent. Let the inner cylinder be at rest, the lowest Reynolds numbers of the outer

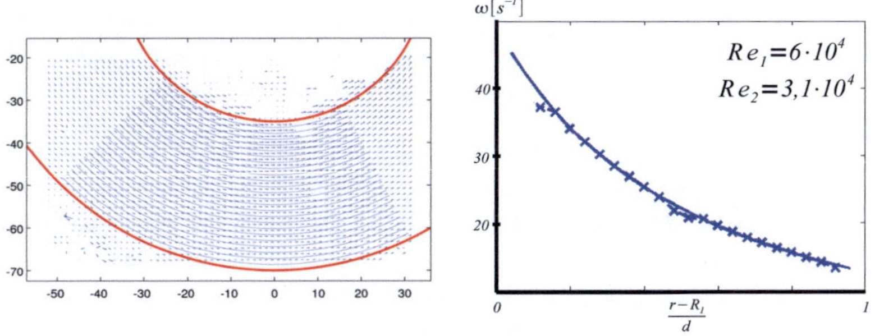

FIGURE 5.1: Left: Results from PIV measurements inside the transparent TC for Reynolds-numbers $Re_1 = 60000, Re_2 = 31000$. The light sheet is adapted in azimuthal-radial plane. The field of view is about $2\pi/5$. The shown measurement is time-averaged over $6.33s$. Right: Radial angular velocity profile out of the PIV measurement averaged over $2\pi/5$.

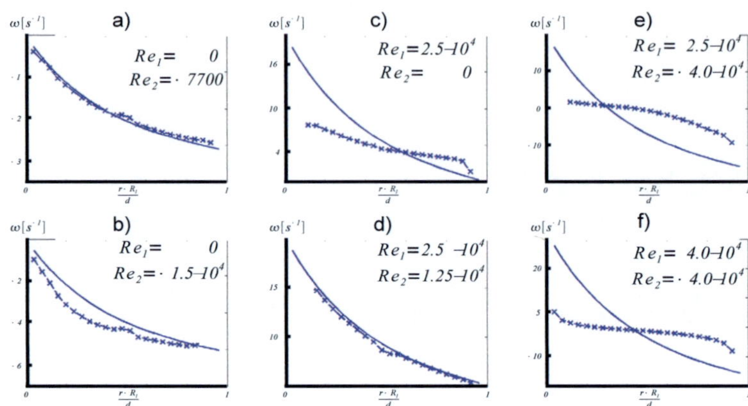

FIGURE 5.2: Angular velocity profiles for different pairs of inner and outer rotation Reynolds numbers. The solid line shows the theoretical profile of laminar Couette flow $\omega(r) = Ar + B/r$, with A and B geometrical factors, where the crosses indicate the measured angular velocities $\omega = u_\varphi/r$. Negative Reynolds numbers indicate clockwise rotation of the corresponding cylinder. a,b) Inner cylinder at rest while outer cylinder rotates with different rotation rates (linear stable flow); c) Inner cylinder rotation only (linear unstable flow); d) Inner and outer cylinder co-rotate at the Rayleigh-criterion ($\eta = \mu^2$); e+f) Counter rotation cases with different inner rotation number (linear unstable flow).

cylinder ($Re_2 = 7700$) shows the same angular velocity profile as the theoretical laminar case. If the outer cylinder is rotating at higher Reynolds numbers the experimental profile obviously differs from the laminar Couette profile, even this is a linear stable flow. For the case of inner rotation one expects a big deviation between the real flow and the Couette profile (see Fig. 5.2c). However, for the case of a co-rotation near the Rayleigh criterion ($\eta = \mu^2$, Fig. 5.2d), the experimentally determined profiles again match with the Couette flow. A further investigation on that apparatus is going to observe the onset of instabilities by crossing the Rayleigh line.

For the cases of counter rotation (see Fig. 5.2f) where $\mu = -0.5$ the angular velocity profiles show a flat radius dependency in the bulk flow and a strong gradient at the boundaries. The boundary layer is not well resolved in this case. The big distinction to the analytic Couette profile shows, that the transport of angular momentum across the gap is increased. In contrast to the flat bulk flow profile for $\mu = -0.5$, we observe a bigger influence of the outer boundary for $\mu = -0.8$ (Fig. 5.2e). At half of the gap width the impact of the gradient at the boundary is strong.

Profiles of the angular velocity are also measured inside the Turbulent Taylor-Couette Cottbus experiment with transparency only in radial direction. A LDV probe is traversed

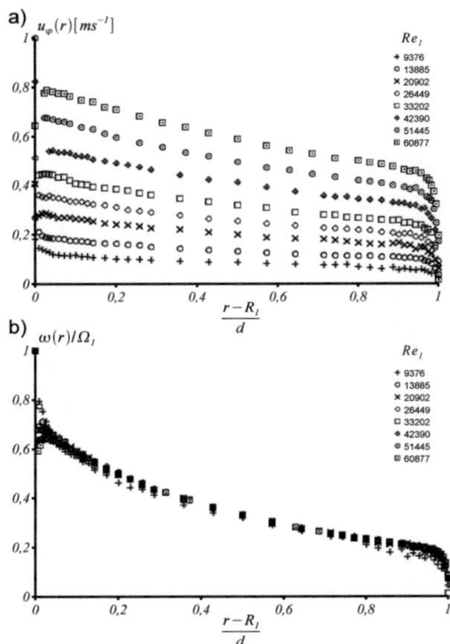

FIGURE 5.3: a) Time averaged azimuthal velocity in dependency on the dimensionless gap width for different Reynolds numbers of the inner cylinder rotation taken by LDA measurement. b) Dimensionless time averaged angular velocity for different Reynolds numbers of the inner cylinder rotation. The values are made dimensionless by the specific angular velocity of the cylinder.

over the whole radius in mid-hight of the experiment for several rotation rates of the inner cylinder. The temporal average of the azimuthal velocity profiles is shown in Fig. 5.3a. The boundary layer at the outer cylinder is resolved in this measurement with several positions, while the layer at the inner boundary is to thin to resolve here. The angular velocity $\omega = u_\varphi/r$ is non-dimensionalized by the rotation speed of the inner cylinder Ω_1. Hence, the dimensionless angular velocity profiles show an identical behavior over the entire range of $Re_1 = 9000$ up to 61000 (Fig. 5.3b). In comparison to the profiles measured with PIV (Fig. 5.2c), this is in good agreement. Further experiments are going to investigate the behavior of angular velocity for even smaller as well as much bigger Reynolds number of the inner rotation and the influence of the outer rotation on the angular velocity.

In both experiments the velocity profiles have been measured using LDV in the TTCC experiment and PIV in the TvTC experiment. Due to the axial transparency of the TvTCC the measurements reveal a much better view. As PIV is possible though the upper plate without the problems of the light defraction of the outer cylinder wall azimuthal-radial velocity fields can be measured. In the following chapter we are going to apply this technique in more detail by the use of several planes at different heights.

5.2 Velocity field measurements: flow structures and plumes

With the use of a high-resolution particle image velocimetry velocity profiles, the wind Reynolds number and characteristics of turbulent plumes in Taylor-Couette flow for the radius ratio of 0.5 and shear Reynolds number of up to $6.2 \cdot 10^4$ are measured. This work was a cooperation of the group at BTU Cottbus and the University of Twente, Enschede funded by the European High-Performance Infrastructures in Turbulence (EuHIT). The results were published in Journal of Fluid Mechanics [36] and this sections will give an overview on these findings.

In this study we are interested to measure the processes of the angular momentum transport through the gap for the different driving scenarios. The roll structures in Taylor-Couette play an important role driving the momentum transport. Even for the ultimate regime of turbulence and counter rotating cylinders ($\mu < 0$) turbulent Taylor-Vortices are observed (see Chapter 6) while for pure inner cylinder rotation the identification of a large circulation was not detectable. In section 5.2.2.3 the roll structures are characterized in the highly-resolved PIV at pure inner cylinder rotation ($\mu = 0$) as well as at the torque maximum ($\mu = -0.2$). A key advantage of the wide gap system ($\eta = 0.5$) is that the torque observations revealed, that the classical turbulent regime lasts up to a shear Reynolds number of $Re_S = 8 \cdot 10^4$ while for wider gaps (i.e. $\eta = 0.71$) the transition to the ultimate regime takes place much earlier at about $Re_S = 1.5 \cdot 10^4$ ([17, 27]). As the scaling of the angular momentum transport changes due to the change of the "wind" in the gap with the transition to he ultimate regime the wide gap is well suited to study the effects of the velocity profiles, statistics, and roll structures in the two different regimes. The measure of the strength if the secondary flows u_r and u_z is named the "wind" in the gap of the cylinders and can be determined by the wind Reynolds number. To quantify the wind Reynolds number from the PIV the standard deviation of the radial velocity $\sigma(u_r)$ is used:

$$Re_w = \sigma(u_r)d/\nu \tag{5.1}$$

The wind Reynolds number dependency on the shear Reynolds number (or the Taylor number: $Ta = r_a^8 r_g^{-8} \cdot Re_S^2$, which relates to $Ta \sim 1.6 \cdot Re_S^2$ for $\eta = 0.5$) and the ratio of angular velocities μ helps to understand the scaling of angular momentum transport and can be compared to theoretical predictions (section 5.2.2.2).

Using the time-resolved azimuthal and radial velocity fields at different heights the roll structures that generate strong radial and axial flow, and the turbulent plumes that emanate from either the inner or outer cylinder and their connection is analysed and reported in van der Veen et al. [36]. In addition the influence of the plumes and large scale circulation to the logarithmic nature of the velocity profiles is discussed there as well.

5.2.1 Setup & explored parameter space

The experiments were carried out in the Top view Taylor-Couette facility, using the inner cylinder radii of $r_i = 35.0 \pm 0.2$mm , leading to the radius ratio of 0.5. As the aspect ratio is sufficiently large the end walls do not significantly affect the velocity in the bulk [31, 37]. The cylindricities of the cylinders that were used are $0.4mm$ and $0.3mm$ for the inner and outer cylinder, respectively. The maximum rotation rates are 5 Hz for both the inner and outer cylinder.

A high-resolution PIV camera (LaVision Imager sCMOS) with a resolution of 2560 × 2160 pixels and a frame rate of 50 Hz is installed above the top end plate, pointing downwards. Water is used as the working fluid ($20°$C, $\nu = 1.0e - 6m^2/s$). The water contains fluorescent particles (Dantec Dynamics, PMMA-RhB, 1-20 μm) with a maximum Stokes number of St $= \tau_p/\tau_\eta \approx 10^{-4} \ll 1$, which means that they faithfully follow the flow and can be considered as tracer particles. The flow is illuminated by a horizontal light sheet from a high-powered pulsed Nd:YLF dual cavity laser (Litron LDY303HE). Because of the high-resolution PIV camera, very high resolution measurements of the flow fields can be achieved. The imaging of the full width of the gap combined with a vector grid of 16 × 16 pixels with 50% overlap results in a velocity vector spacing of 0.13 mm. The PIV system is operated in dual frame mode, allowing for an interframe time Δt smaller than the inverse frame rate. The PIV image pairs are processed using LaVision DaVis software, after which the flow fields are transformed to the radial velocity $u_r(\theta, r, z, t)$ and the azimuthal velocity $u_\theta(\theta, r, z, t)$.

The driving parameters were varied in two sets. First for pure inner cylinder rotation ($\mu = 0$) and shear Reynolds numbers between Re$_S = 6.0 \cdot 10^3$ to $6.2 \cdot 10^4$ (corresponding to Taylor numbers of Ta $= 5.8 \cdot 10^7$ and Ta $= 6.2 \cdot 10^9$) only at mid-height. Each of these measurements consists of 10000 PIV image pairs with a sampling rate of 25Hz. The second set was to reveal the tree dimensional roll structure. The height of the laser sheet was traversed to 13 different heights with 7.5 mm spacing, 5000 image pairs of the velocity field were recorded for Re$_S = 5 \cdot 10^4$ for both pure inner cylinder rotation and optimal [?] counter-rotation $\mu = -0.2$.

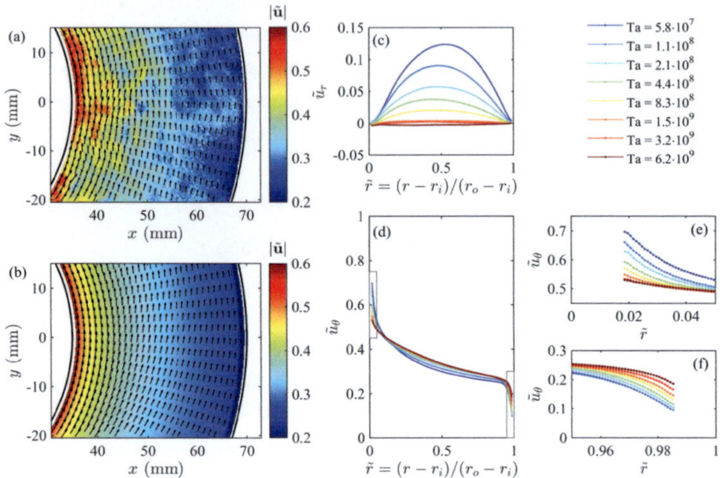

FIGURE 5.4: Overview of the flow profiles for varying Taylor number with pure inner cylinder rotation $\mu = 0$ at mid-height $h/L = 0.5$. (a) Snapshot of the flow field for $Ta = 6.2 \cdot 10^9$. The colours and lengths of the arrows indicate the norm of the velocity $|\tilde{\mathbf{u}}|$. (b) Averaged flow field over 10000 PIV image pairs at $Ta = 6.2 \cdot 10^9$. (c) Radial velocity profiles across the gap of the TC apparatus, normalised by the inner cylinder velocity. For lower Taylor numbers there is still a strong radial flow, which can be attributed to the presence of Taylor vortices. (d) Azimuthal velocity profiles normalised by the inner cylinder velocity. For increasing Taylor number, the profiles become flatter and the boundary layers steeper. (e) and (f) Magnification of the azimuthal velocity profiles in (d) close to the cylinders. Figure from van der Veen et al. [36].

5.2.2 Results

5.2.2.1 Azimuthal and angular velocity profiles

From the measurements instantaneous flow fields are gained, an example is given in Figure 5.4(a), these lead to the time averaged fields shown in Figure 5.4(b). Normalizing the radial velocity $\tilde{u}_r = u_r/u_\theta(R_1)$ and plotted versus the dimensionless gap position $\tilde{r} = (r - R_1)/d$ (Fig. 5.4(c) shows that at mid-height a significant mean radial flow is existing for the lower shear Reynolds numbers. These Taylor vortices disappear for outer cylinder at rest and higher Reynolds numbers [20]. As the shear Reynolds number increases, the azimuthal velocity profile is getting flatter in the bulk flow (figure 5.4(d-f)), while the strong deviation of the boundary layers at inner compared to outer wall, caused by the strong curvature is observable for the high Reynolds numbers.

To distinguish the asymetric boundary layers, in figure 5.5 the profiles are given from wall to centre of the gap while the location is turned into the normalised distance from

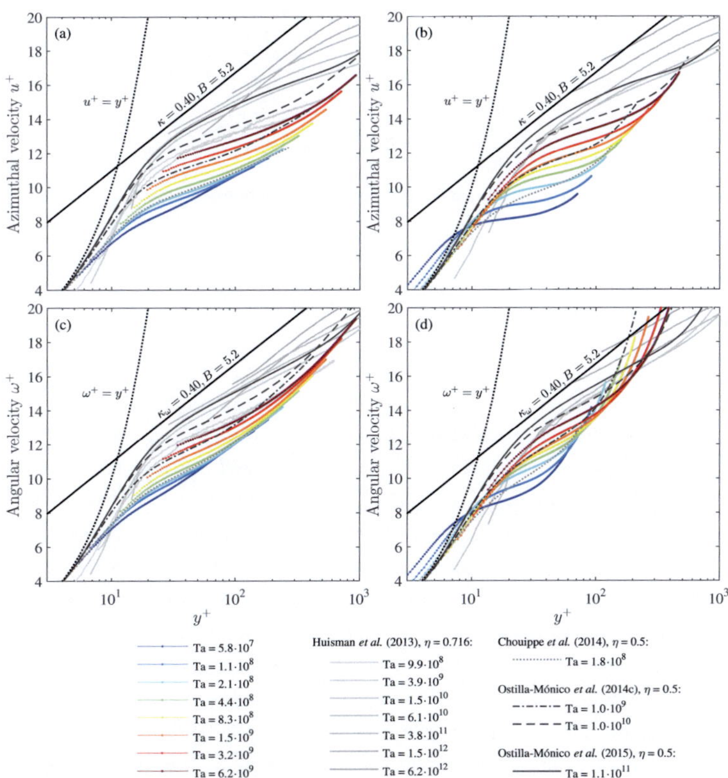

FIGURE 5.5: Velocity profiles for inner cylinder rotation $\mu = 0$ and varying Taylor number. See main text for the definitions of u^+, ω^+ and y^+. (a) Azimuthal velocity profiles near the inner cylinder ($\tilde{r} \in [0, 1/2]$). (b) Azimuthal velocity profiles near the outer cylinder ($\tilde{r} \in [1/2, 1]$). (c) Angular velocity profiles near the inner cylinder ($\tilde{r} \in [0, 1/2]$). (d) Angular velocity profiles near the outer cylinder ($\tilde{r} \in [1/2, 1]$). All figures include the logarithmic law of the wall from Prandtl and von Kármán $u^+ = 1/\kappa \ln y^+ + B$ with the typical values of $\kappa = 0.40$ and $B = 5.2$ (see Marusic et al. [23] and references therein), the viscous sublayer $u^+ = y^+$, DNS data from Chouippe et al. [7], Ostilla-Mónico et al. [28] and Ostilla-Mónico et al. [27] at $\eta = 0.5$ and measurement data from ?] at $\eta = 0.716$. Figure from van der Veen et al. [36].

the wall y^+ from both the inner and outer cylinders respectively. Also the azimuthal velocity is computed into the 'inner' coordinates as u^+ ((a) and (b)) and angular velocity ω^+ ((c) and (d)). The inner coordinates are defined as follows:

$$y^+ = \frac{r - R_1}{\delta_{\nu,1}} \tag{5.2}$$

$$u^+ = \frac{u_\theta(R_1) - u_\theta(r)}{u_{\tau,1}} \tag{5.3}$$

$$\omega^+ = \frac{\omega(R_1) - \omega(r)}{u_{\tau,1}/R_1}; \tag{5.4}$$

The same parameters for the outer cylinder are defined as:

$$y^+ = \frac{R_2 - r}{\delta_{\nu,2}} \tag{5.5}$$

$$u^+ = \frac{u_\theta(r) - u_\theta(R_2)}{u_{\tau,2}} \tag{5.6}$$

$$\omega^+ = \frac{\omega(r) - \omega(R_2)}{u_{\tau,2}/R_2} \tag{5.7}$$

with the viscous length scale $\delta_{\nu,1i} = \nu/u_{\tau,i}$, the friction velocity $u_{\tau,i} = \sqrt{\tau_{w,i}/\rho}$ containing the wall shear stress $\tau_{w,i} = T/2\pi R_i^2 L$ and the torque T. As a consequence $\delta_{\nu,1}/\delta_{\nu,2} = \eta$, $u_{\tau,1}/u_{\tau,2} = 1/\eta$ and $\tau_{w,1}/\tau_{w,2} = 1/\eta^2$. The torque values are from Chapter 4 and also Merbold et al. [24]. The torque from these measurements are globally measured while the velocity measurements are local. The local torque can be different from the global torque due to the large-scale roll structures, leading to an imperfect matching with the viscous sublayer $u^+ = y^+$ and $\omega^+ = y^+$, especially visible in Fig. 5.5(b) and (d). By increasing the shear Reynolds number the boundary layer profiles tend to Prandtl-von Kármán log-law, while at these Reynolds numbers the turbulent log layer is not fully developed. A good agreement is given with the DNS performed from Chouippe et al. [7] at $Re_S = 1.1 \cdot 10^4$, Ostilla-Mónico et al. [28] at $Re_S = 2.4 \cdot 10^4$ and $Re_S = 7.5 \cdot 10^4$ and Ostilla-Mónico et al. [29] at $Re_S = 2.6 \cdot 10^5$. To compare the profiles also the results by ?] for $\eta = 0.716$ are shown.

Grossmann et al. [16] argued that the angular velocity profiles ω^+ are closer to the logarithmic law than the azimuthal velocity profiles u^+, this can be confirmed by the measurements taken (Fig. 5.5 Compare (a) to (c) and (b) to (d)), especially for the stronger curvature effect of the small radius ratio.

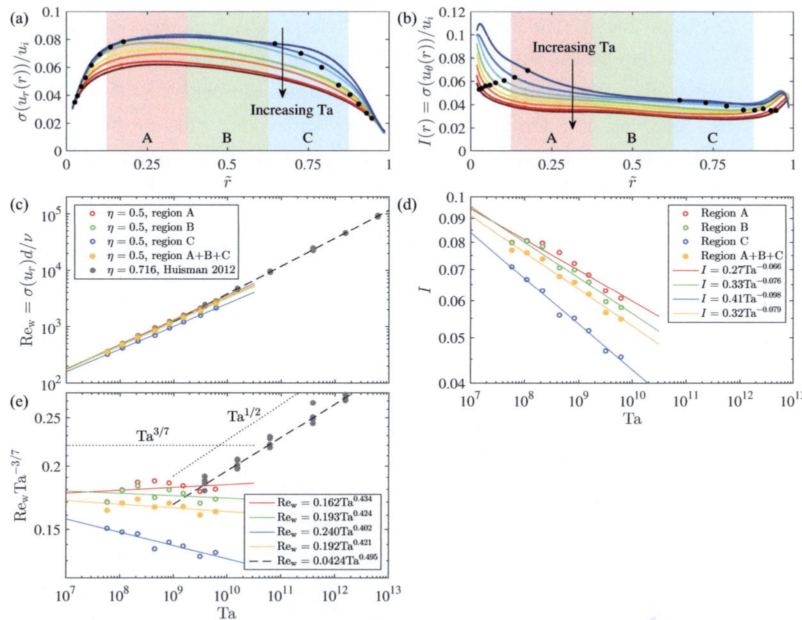

FIGURE 5.6: The scaling of the wind Reynolds number Re_w and turbulence intensity I with Taylor number for inner cylinder rotation $\mu = 0$. (a,b) The standard deviation of the radial velocity (a) and azimuthal velocity (b) over 10000 frames (400 s) and the azimuthal direction, normalised by the inner cylinder velocity, as a function of the normalised radial position. This corresponds to the turbulence intensity I in the case of the azimuthal velocity. See the legend in figure 5.4 or 5.5 for the values of the Taylor / shear Reynolds numbers. The three areas A ($\tilde{r} \in [1/8, 3/8]$), B ($\tilde{r} \in [3/8, 5/8]$) and C ($\tilde{r} \in [5/8, 7/8]$) indicate regions over which σ is averaged for figures (c-e). The black dots represent the position of the start of the outer layers at $y^+ = 50$. (c) Re_w versus Ta averaged over $\tilde{r}$-values corresponding to regions A, B, C and all three combined ($\tilde{r} \in [1/8, 7/8]$). In addition, data from Huisman et al. [17] at $\eta = 0.716$ with their fit are shown, which are fully in the ultimate turbulent regime. This data uses the equivalent to area B for averaging. (d) Turbulence intensity I versus Ta averaged over the same regions as before. The fits are $I = 0.27Ta^{-0.066}$ for area A, $I = 0.33Ta^{-0.076}$ for area B, $I = 0.41Ta^{-0.098}$ for area C and $I = 0.32Ta^{-0.079}$ for the three areas combined. (e) Data in (c) compensated by $Ta^{3/7}$. The fits are $Re_w = 0.162Ta^{0.434}$ for area A, $Re_w = 0.193Ta^{0.424}$ for area B, $Re_w = 0.240Ta^{0.402}$ for area C and $Re_w = 0.192Ta^{0.421}$ for the three areas combined. Figure from van der Veen et al. [36].

5.2.2.2   Wind Reynolds number and turbulence intensity

The theory by Grossmann and Lohse [14] and Eckhardt et al. [10] derives the analogy of Rayleigh-Bénard convection and Taylor-Couette flow. For the wind Reynolds number it can be predicted, that it scales with the Taylor/Rayleigh number $Re_w \propto Ta^{3/7}$, $Rew \propto Re_s^{6/7}$ in the classical turbulent regime. When the bulk as well as the boundary layers are turbulent - in the ultimate regime - this scaling should behave as $Re_w \propto Ta^{1/2}$, $Re_w \propto Re_S$ following [15]. For Taylor-Couette flow with radius ratio $\eta = 0.716$ this has been experimentally confirmed by Huisman et al. [17] up to $Ta = 6.2 \cdot 10^{12}$, respectively $Re_S = 2.3 \cdot 10^6$.

For the present measurements the highest Reynolds number reached is $Re_S = 6.2 \cdot 10^4$ the flow is in the classical turbulent regime. Thus, the theoretical prediction of 3/7-scaling is expected to be hold. This scaling has not been confirmed in Taylor-Couette flow so far but is known for Rayleigh-Bénard convection Grossmann and Lohse [14].

The wind Reynolds number is measured in terms of a radial average of the standard deviation of the radial velocity component in time and azimuthal direction $\sigma(u_r(r,\theta,t))$ calculated from the velocity field $u_r(\theta,r,z,t)$ across the gap, see Figure 5.6. The dependence of $\sigma(u_r)$ on the relative gap position $\tilde{r}$ shows a strong asymetry (Figure 5.6(a)), where a maximum is observered at about one quarter of the gap width away from the inner cylinder. The huge difference in the curvature of the inner and outer cylinder for radius ratio $\eta = 0.5$ causes these asymetries in the flow field. To compute the wind Reynolds number Re_w, the choice of the radial region, where $\sigma(u_r(r,\theta,t))$ is averaged, has an impact. As the bulk transport is of interest it is reasonable excluding the boundary layers. Figure 5.6(c) shows the behaviour of $Re_w(Ta)$ with different areas (A; B; C; A+B+C) of averaging.

A clear power law scaling is observed with an exponent ranging between 0.402 up to 0.434 (depending on the average-region chosen) as depicted in Figure 5.6(e). It is noteworthy that this is in agreement with the prediction for the classical turbulent scaling $3/7 = 0.429$ mentioned earlier. Thus, the data also agrees with the findings of Huisman et al. [17] for narrow gap, while the difference in radius ratio would not presume this agreement.

5.2.2.3   Roll structures

The roll structures, the turbulent plumes and the resulting logarithmic boundary layer velocity profiles are three features of Taylor-Couette flow, which are closely related to each other. The roll structures in classical TC flow happen already in laminar flow

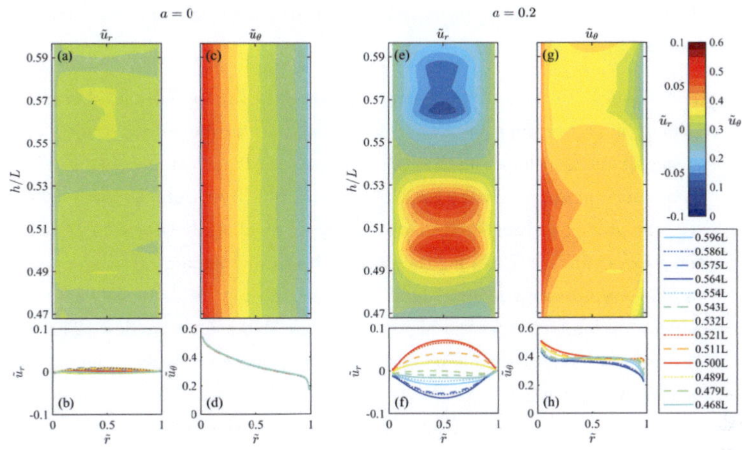

FIGURE 5.7: Dependence of flow profiles on axial position for rotation ratios $a = 0$ (a-d) and $a = 0.2$ (e-h) for $Ta = 4.2 \cdot 10^9$. The normalised radial velocity is $\tilde{u}_r = u_r(r)/(u_i - u_o)$ and the azimuthal velocity is $\tilde{u}_\theta = (u_\theta(r) - u_o)/(u_i - u_o)$. The data is represented in a colour map (a,c,e,g) and as profiles in (b,d,f,h). Figure from van der Veen et al. [36].

and are still a feature of the turbulent TC flow. In the present measurements the roll structures are visible in the radial flow at different heights. Thus, flow profiles are measured at several heights for the case of outer cylinder at rest ($\mu = -a = 0$) and the case of rotation ratio of torque maximum ($mu = -a = -0.2$) representing the roll structures in Figure 5.7. This happened at the shear Reynolds number of $Re_S = 5 \cdot 10^4$, corresponding to $Ta = 4.2 \cdot 10^9$. For pure inner cylinder rotation the roll structures can not be identified, while a slight radial velocity component gives a small hint, that remnants of vortices are in the flow. In comparison to that, for the case of torque maximum, apparently the radial velocity shows the existence of strong roll structures. The fact that there are strong rolls at the maximum torque counter rotation will be discussed in further measurements (Chapter 6, 7)

At the mid gap ($h/L = 0.5$ up to $h/L = 0.52$) the radial velocity shows a prominent outflow, given in figure 5.7(e). while at about $h/L = 0.57$ to 0.59 a strong inflow is observed. In beween a large-scale circulation happens with a length of about $\Delta h/L = 0.05$, respectively of about 1 gap width as the aspect ratio is $\Gamma = 20$. However, the central outflow lays a bit of the mid height of the system. But, small asymmetry has been already observed in Taylor-Couette flow in different setups Huisman et al. [18], van der Veen et al. [36] and others. Interestingly the outflow has two maxima at heights $0.52L$, corresponding to the bottom of a roll and the second maxima at $0.50L$, corresponding to the top of the neighbouring roll. In between these maxima the radial velocity has a local minimum.

The azimuthal velocity component is depicted in Figure 5.7(g,h). Obviously the radial velocity advects azimuthal velocity from the inner to the outer cylinder. At the location of strong outflow the azimuthal velocity profiles are deflected to higher values in near of to the inner cylinder. At the location of radial flow from outer to inner cylinder (negative u_r), the profiles are deflected to smaller values of azimuthal velocity, especially in near of the outer cylinder. Thus, for the case of the torque maxima ($a = -\mu = 0.2$) the strong radial velocity component u_r and its axial dependency let us identify a strong large-scale-circulation happening in the highly turbulent Taylor-Couette flow. This radial velocity also advects the azimuthal velocity across the gap and changes the azimuthal velocity profiles along the height of the system. In the expressed Study of van der Veen et al. [36] also the dependency of turbulent plumes and their effect on the velocity profiles is discusses as well as the logarithmic behaviour of the flow close to the cylinder walls.

Chapter 6

Flow Visualisation

6.1 Flow structures and its behaviour

In the present chapter the flow patterns in turbulent Taylor-Couette flow are studied. The torque acting to the inner cylinder has been measured and discussed in section 4.1. Here we discuss the flow behaviour. For a constant shear Reynolds number of 5000 and 25000 the ratio of angular velocities has been varied. An axial oscillation of the Vortices has been observed for counter rotation. Also one can observe that the flow turns it direction at the inflow regions of the large scale circulation. By our experimental observations of the flow patterns the resulting torques are discussed.

6.1.1 Space-Time diagrams using visualisation techniques

In the present section light visualization of the flow patterns for different flow states are described and analysed. For this investigation the TvTCC is used and filled with silicone oil of a high viscosity ($\nu = 20.753 \cdot 10^{-6} m^2/s$). In addition aluminium flake particles are added to make the flow behaviour visible.

For capturing images and videos of the flow patterns a Camera Canon EOS 1100D is used. The aluminium particles are small flakes of approximately $5 \mu m$ size. They follow the flow such that vortices can be identified by the light they reflect. In Fig. 6.1 (left) one can see the visualization of laminar Taylor vortices ($Re_S = 450$). For the case of inward or outward welling flow the flakes are oriented that they reflect less light. For the case of strong axial flow (upwards and downwards) the flakes show there flat site and reflect the light strongly. Thus, structures can be identified. In the case of $Re_S = 1000$ the flow becomes an additional axial instability and shows the so called Wavy Vortex flow.

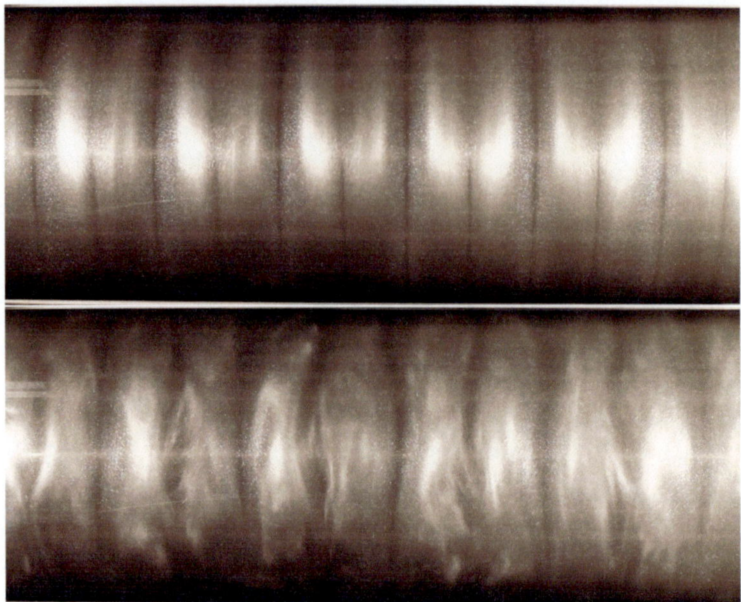

FIGURE 6.1: Visualization of (Left) the Taylor-Vortex flow at $Re_S = 450$ and (Right) Wavy vortex flow at $Re_S = 1000$ while the outer cylinder remains at rest.

If the flow becomes additional instabilities or even a turbulent behaviour the particles will make this visible.

In Fig. 6.2 the flow patterns for $Re_S = 5000$ are plotted for the case that (left) the outer cylinder is at rest ($\mu = 0$) and (right) for the inner cylinder at rest ($\mu \to \infty$). For $\mu \to \infty$ the flow is laminar because of its stable stratification of angular momentum. In contrast for $\mu = 0$ turbulent Taylor-Vortices appear. In the right figure the black solid line indicates the position where we perform a deeper analysis of the temporal behaviour of the flow.

To investigate the temporal evolution of such flows images are taken by a camera Canon EOS 1100D. The camera is capturing videos of the fluid flow with 25 frames per second. Now the brightness information at the central position in dependence of the axial position is read out. This brightness information is then plotted in a space-time diagram. To reduce the influences of inhomogeneous illumination and disperse light reflection of surrounding bodies a grey image is taken and subtracted from the other recorded images. In Fig. 6.3 the brightness course of the grey image along the black line of Fig. 6.2 is shown.

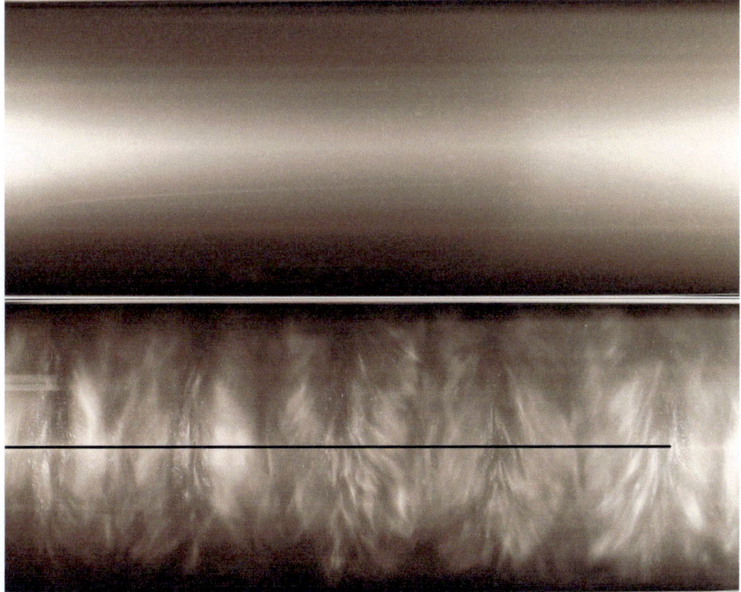

FIGURE 6.2: (Left) Visualization of laminar Couette flow at $Re_S = 5000$ for inner cylinder at rest. (Right) Turbulent Taylor-Vortices for $Re_S = 5000$ and outer cylinder at rest. In a single image it becomes difficult to identify the Taylor Vortices. Watching to the time dependent flow one can observe them better (Fig 6). The black line indicates the position were the brightness information of the videos is plotted into space-time diagrams such as follows.

Due to the positioning of the lights near the end plates, the brightness value is lower in the center of the axial position than of the outer regions. Fig. 6.4 shows the time space diagram of the wavy vortex flow for $Re_S = 1000$. Taylor vortices are visible which appear bright in regions of axial flow direction next to the outer cylinder and dark in regions of radial flow direction, the so called inflow and outflow boundaries. Further an axial oscillation of the Taylor vortices is seen indicating the characteristic azimuthal waves of the Wavy Vortex flow.

6.1.2 Flow patterns for $Re_S = 5000$

As we discussed in section 2 the angular motion transport within the gap has been measured already for a wide range of Reynolds numbers. Especially for $Re_S = 5000$ a very good agreement of numerical calculations and the experimental measured torque is given [24]. As I discuss the visualisation method for this shear Reynolds number within

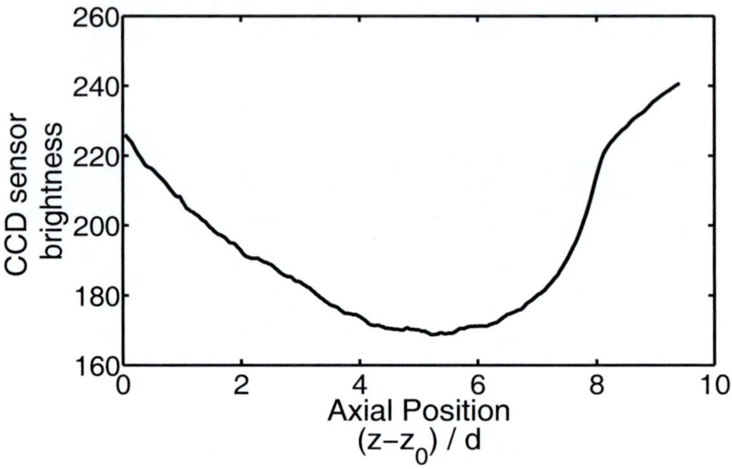

FIGURE 6.3: Brightness course of the grey image along an axial line.

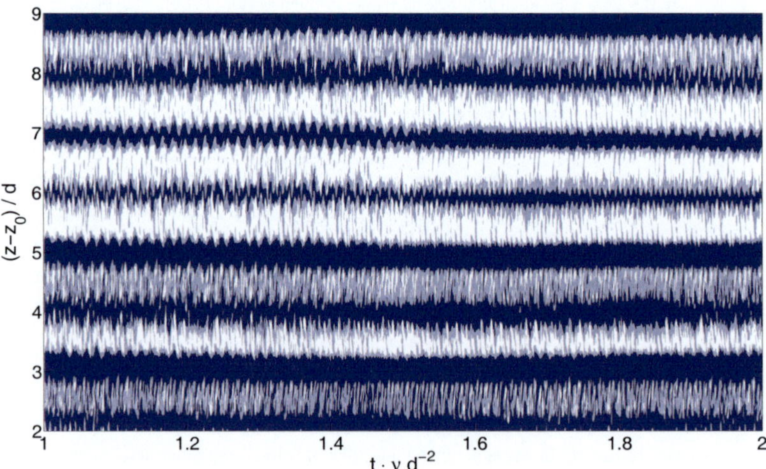

FIGURE 6.4: Space time diagram of a Wavy Vortex flow with $Re_S = 1000$ and the outer cylinder at rest.

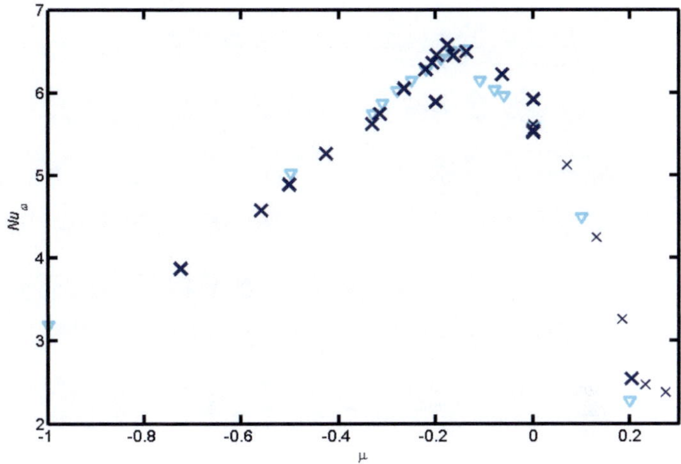

FIGURE 6.5: The quasi-Nusselt number Nu_ω as function of the rotation rate μ at constant shear Reynolds number $Re_S = 5000$. The experimental results (crosses) are compared to numerical simulations (triangles) [24].

this section, the torque is plotted again in Fig. 6.5 in comparison between experimental and numerical findings.

As seen in Fig. 6.5 the angular motion transport becomes larger for a certain counter rotation while it decreases for strong counter rotation. The maximum of the transport is located at $\mu_{max} = -0.20$. Some ideas are given by different authors what process drives this transport phenomenon [2, 37]. In the present investigation we discuss the observed flow structures and their temporal behaviour. The videos captured help us to understand the structures and the space-time plots of the visualization videos show us the temporal behaviour for the different cases. In the following the observed flow patterns for the constant shear Reynolds number of $Re_S = 5000$ is plotted for different values of the ratio of angular velocities. For the case that the outer cylinder is at rest ($\mu = 0$) we observe turbulent Taylor-Vortices. Taylor vortices exist and dominate the patterns (Fig. 6.6) which become strongly turbulent. One can observe this already in a single image (Fig. 6.2 right) compared to laminar Taylor-Vortex flow (Fig. 6.1 left). Also in the temporal behaviour (Fig. 6.6) the turbulent fluctuations are clearly visible compared to a laminar flow such as the Wavy-vortex flow in Fig. 6.4. Other structures or large scale fluid motions are not noticeable. Also a strong difference exists for the case of pure rotation of the outer cylinder. Centrifugal instability does not happen in this situation due to the Rayleigh-criterion. Also the occurrence of turbulent structures

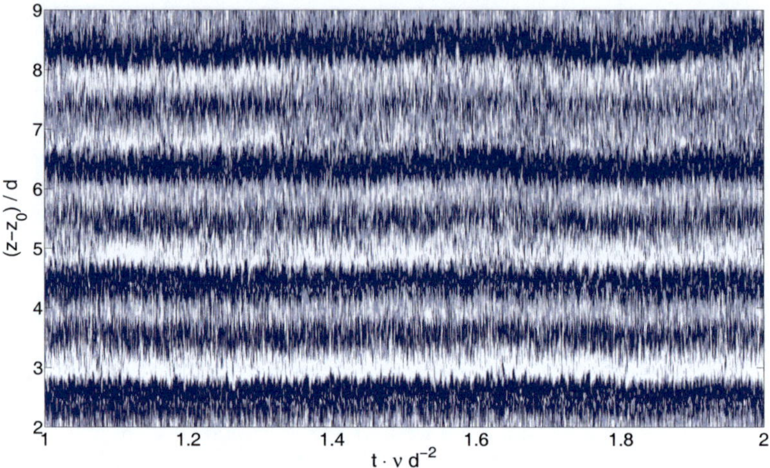

FIGURE 6.6: Space time diagram of regular turbulent Taylor-Couette flow for $Re_S = 5000$, the outer cylinder is at rest ($\mu = 0$).

is suppressed by the stable stratification of angular momentum inside the cylindrical gap. Instabilities driven by the solid end plates can now become important and they do grow into the flow. In the following we discuss the flow between counter rotating cylinders. Here the effects of both should play a role. When the influence of the inner cylinder rotation is dominant (μ close to 0) the centrifugal instability and turbulence drives the angular motion. If the influence of the outer cylinder becomes important the stabilizing effect suppresses centrifugal instabilities and calms the turbulent flow. One would mention that decreasing the ratio of angular velocities μ the transported angular motion should decrease as well. As we observe an increase up to the maximum of torque at $\mu_{max} = -0.20$ interesting flow features happen.

As described before, when the ratio of angular velocities is decreased slightly to $\mu = -0.10$, the flow remains in a similar situation. Turbulent Taylor Vortices are also observed in that case. Fluctuations of the flow grow to the cost of Taylor vortices. The flow pattern however does not change in principle (Fig. 6.7).

According to Fig. 6.8 a further decrease of the ratio of angular velocities to $\mu = -0.15$ modifies the flow considerable. The Taylor vortices are deformed in the axial direction to a spike-like structure with a fixed frequency of $0.19Hz$. This frequency is much smaller than the rotation rates of the cylinders ($n_1 = 9.81Hz, n_2 = 1.32Hz$). The axial amplitude of this oscillation is of about half gap width. However the flow is strongly

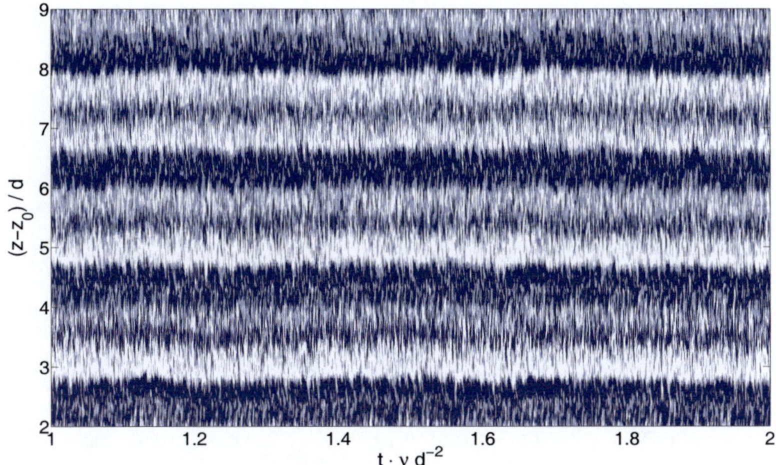

FIGURE 6.7: Space time diagram of slight counter rotating turbulent Taylor-Couette flow for $Re_S = 5000$ and $\mu = -0.1$.

dominated by turbulent fluctuations. In addition one can observe that all the aluminium flakes in the flow do move the azimuthally direction of the inner cylinder. Hence, the neutral azimuthal velocity position must be very close at the outer cylinder.

The flow pattern close to the torque maximum for $\mu = -0.20$ is pictured in Fig. 6.9. The axial oscillating structures are strengthened with an increased amplitude around one times the gap width while the frequency is decreased to $0.14Hz$ (Rotation rates are $n_1 = 8.44Hz, n_2 = 1.69Hz$). The flow is still turbulent at all axial positions and dominated by these oscillating Taylor-Vortices. Using a parallel observation of front and back of the experimental apparatus one can see the global structure. We identify this as an azimuthal wave with mode $m = 1$ while the azimuthal form of the wave is not sinusoidal as for Wavy-Vortex flow. It is more identified to be similar to a zigzag function. Another change occurring for this flow is that at the inflow regions of the vortices the velocity does decelerate to rest in azimuthal direction. Still the particles are not moving in the direction of the outer cylinder. Thus, the stabilizing effect of the outer cylinder rotation has not yet an influence to the flow.

As we pass through the torque maximum and go to stronger counter rotation $\mu = -0.25$ the oscillating vortices are still observable. But the axial wave structure is weakened. While at the inflow regions of the vortices the flow turns around and the particles are accelerated in the direction of outer cylinder rotation. The outer cylinder rotation

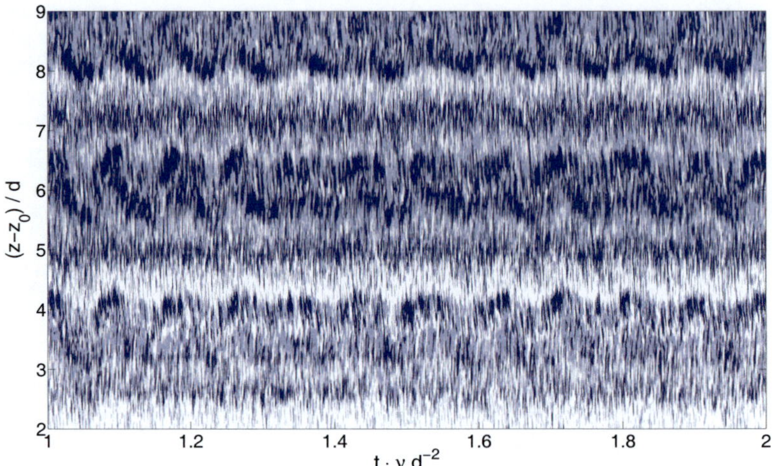

FIGURE 6.8: Space time diagram of slight counter rotating turbulent Taylor-Couette flow for $Re_S = 5000$ and $\mu = -0.15$.

stabilizes the flow and turbulent fluctuations are decreased. At the outflow regions turbulent flow is carried from the inner cylinder outwards to the outer cylinder with a high velocity. The frequency of the axial motion is decreased by a factor of two while the amplitude remains to be of about one gap width.

In Fig. 6.11 the flow with $\mu = -0.30$ is displayed. Inside the turbulence the axial oscillating structure remains weak. The oscillation also becomes irregular and a frequency or amplitude cannot be identified by our visualization. Also it is difficult to identify the vortices by an axial motion of the particles. One can identify them by seeing the turn of azimuthal velocity depending on the axial position. Here the domain where the outer cylinder rotation accelerates the fluid in its direction is enhanced and has a stabilizing effect to the flow. The strong vortices are surely beyond the visible fluid layer of about $2mm$ but they become suppressed in this case. This leads to the decrease of the torque. For very strong counter rotation ($\mu = -1.00$) one cannot identify any vortex like structure in the flow. The visible fluid moves turbulent in the direction of the outer cylinder (Fig. 6.13). This closes our observations on the fluid structures for the measured torques in Fig. 6.5.

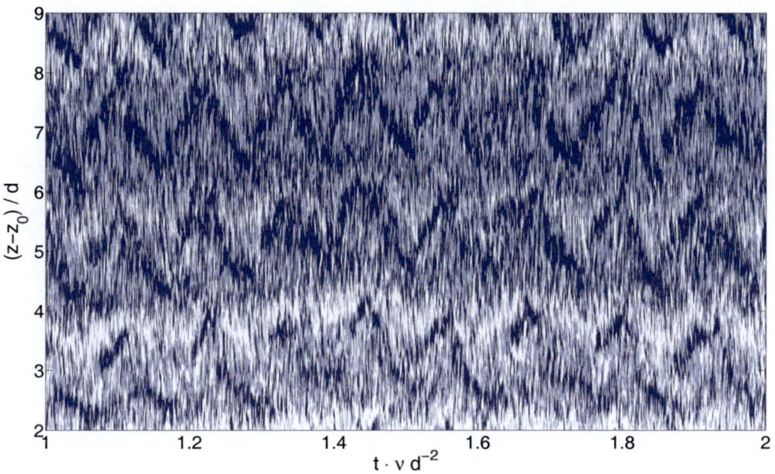

FIGURE 6.9: Space time diagram of slight counter rotating turbulent Taylor-Couette flow for $Re_S = 5000$ and $\mu = -0.20$.

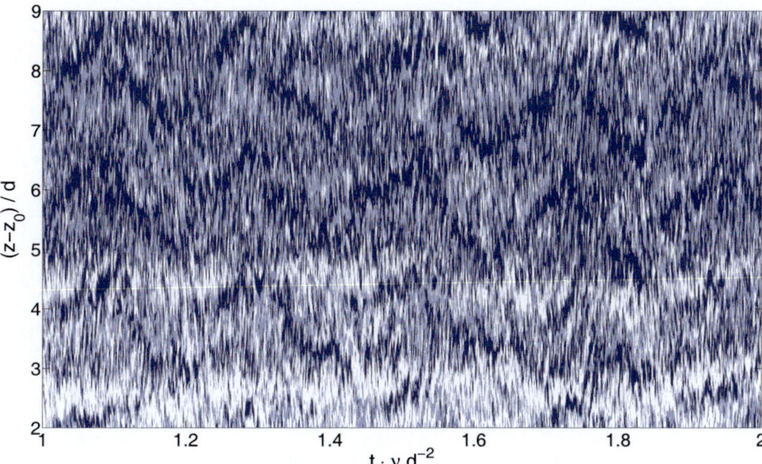

FIGURE 6.10: Space time diagram of slight counter rotating turbulent Taylor-Couette flow for $Re_S = 5000$ and $\mu = -0.25$.

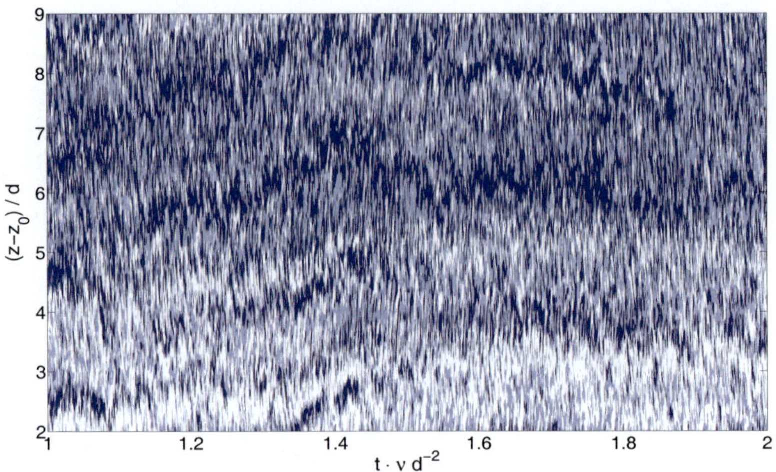

FIGURE 6.11: Space time diagram of slight counter rotating turbulent Taylor-Couette flow for $Re_S = 5000$ and $\mu = -0.30$.

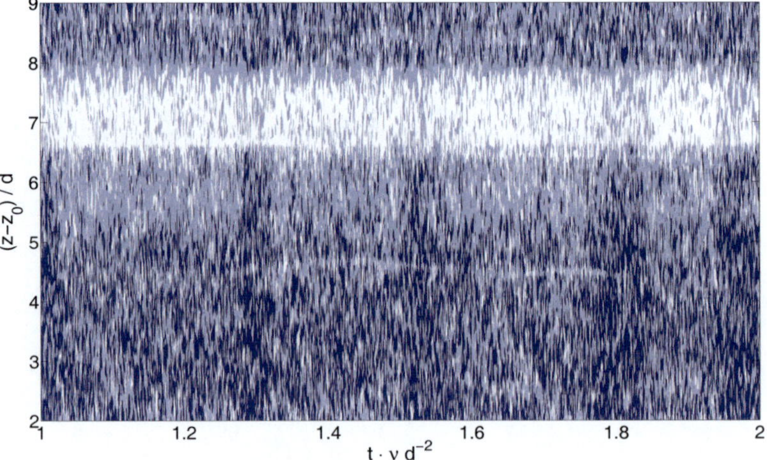

FIGURE 6.12: Space time diagram of slight counter rotating turbulent Taylor-Couette flow for $Re_S = 5000$ and $\mu = -0.50$.

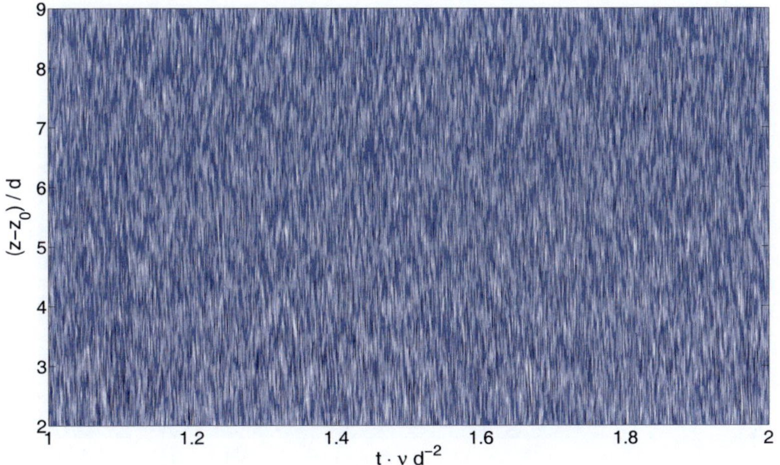

FIGURE 6.13: Space time diagram of slight counter rotating turbulent Taylor-Couette flow for $Re_S = 5000$ and $\mu = -1.00$.

6.1.3   Flow Patterns for $Re_S = 26000$ and 52000

The Visualisation method described for the shear Reynolds number of $Re_S = 5000$ was performed for higher Reynolds numbers. A systematic analysis has been done also for $Re_S = 26000$ and 52000. For both shear Reynolds numbers also the ratio of angular velocities was varied from pure inner rotation ($\mu = 0$) up to very strong counter rotation ($\mu = -0.50$).

It is noteworthy, that the case of $Re_S = 52000$ is close to the analysis given in chapter 5 5.2. The flow behaviour observed for the case of $Re_S = 26000$ is the same as for $Re_s = 52000$. Thus, for brevity, only the space-time diagrams for the large shear Reynolds number is depicted in this chapter, while the flow visualisation of the lower one is depicted in Appendix A.

Also for these shear Reynolds numbers another gray image study was performed to get the sensor brightness undisturbed from the flow and the same procedure is applied. The cameras view did change. Hence, a different field of view is given, while an accurate detection of the location of mid height is performed and used to place the origin of the axial coordinate. The videos were performed for a longer time of measurements. In Figures 6.14 to 6.20 the first fragment of these visualisation is shown. The following fragments are attached in Appendix A.

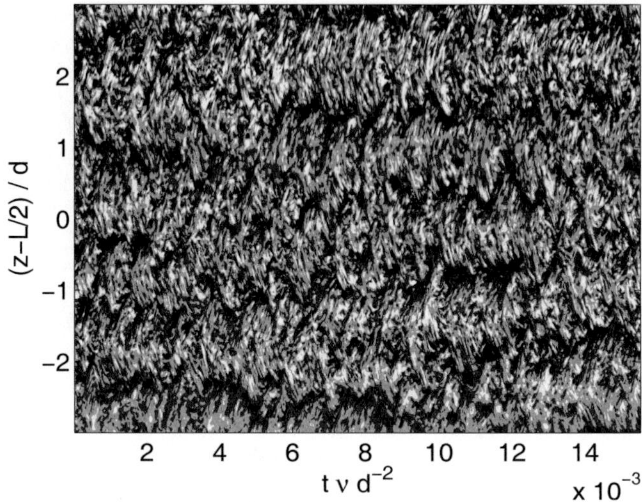

FIGURE 6.14: Space time diagram of turbulent Taylor-Couette flow for $Re_S = 52000$ and $\mu = 0.00$. The diagram shows the first fragment of the measured data, the following data are given in Appendix A, Figures A.1, A.2, A.3.

In the case of pure inner cylinder rotation a strong difference of the flow behaviour is observable. In Figure 6.6 for $Re_S = 5000$ the Taylor vortices are very prominent visible in the space time diagram. As the shear Reynolds number increases the Taylor vortices disappear in the visualisation (Figure 6.14). A turbulent flow is observable but no large scale circulation fix in space can be identified. Obviously the featureless turbulence does dominate the flow.

As the outer cylinder rotation is directed in the other direction, the flow behaviour changes drastically. Figure Figure 6.15 shows the space-time behaviour for the case of $\mu = -0.10$. Obviously, the counter rotation increases the strength of the Taylor vortices. The flow also is turbulent, but the large-scale circulation has set in and creates a local dominant axial flow at the outer cylinder.

As the counter rotation becomes stronger, $\mu = -0.15, -0.20$, the turbulent Taylor vortices become more dominant (Figures 6.16, 6.17). The inclination in the space time plot becomes higher for the positive and negative axial flow. In the latter case the torque maximum is reached. In the space time diagram of the flow visualisation for $Re_S = 52000, \mu = -0.20$ an axial inconsistency of the large-scale circulation is observed. A clear oscillation like a wavy behaviour can not be identified, but an axial shift is given. This can be mostly seen at the two vortices close to the systems mid-height.

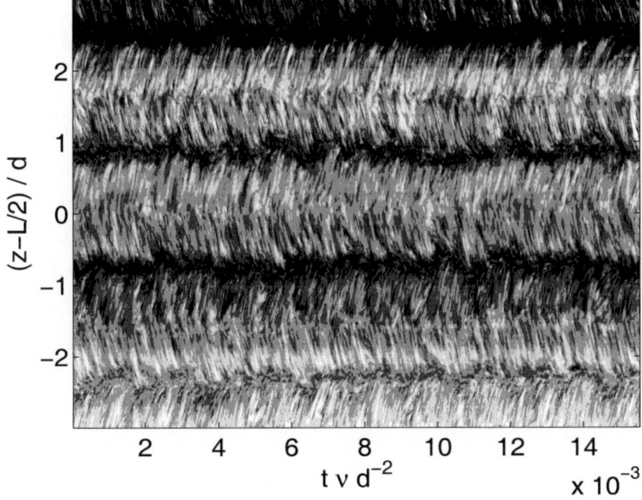

FIGURE 6.15: Space time diagram of turbulent Taylor-Couette flow for $Re_S = 52000$ and $\mu = -0.10$. The diagram shows the first fragment of the measured data, the following data are given in Appendix A, Figures A.4, A.5, A.6.

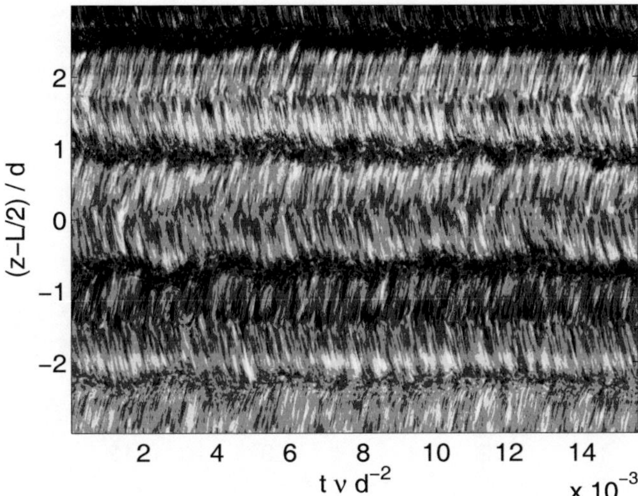

FIGURE 6.16: Space time diagram of turbulent Taylor-Couette flow for $Re_S = 52000$ and $\mu = -0.15$. The diagram shows the first fragment of the measured data, the following data are given in Appendix A, Figures A.7, A.8, A.9.

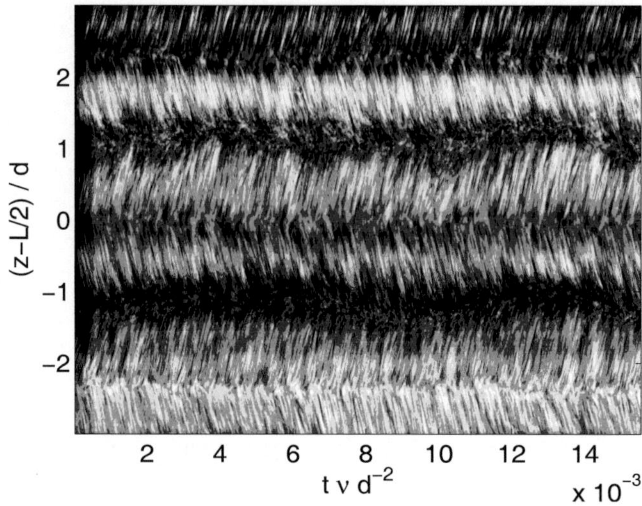

FIGURE 6.17: Space time diagram turbulent Taylor-Couette flow for $Re_S = 52000$ and $\mu = -0.20$. The diagram shows the first fragment of the measured data, the following data are given in Appendix A, Figures A.10, A.11, A.12.

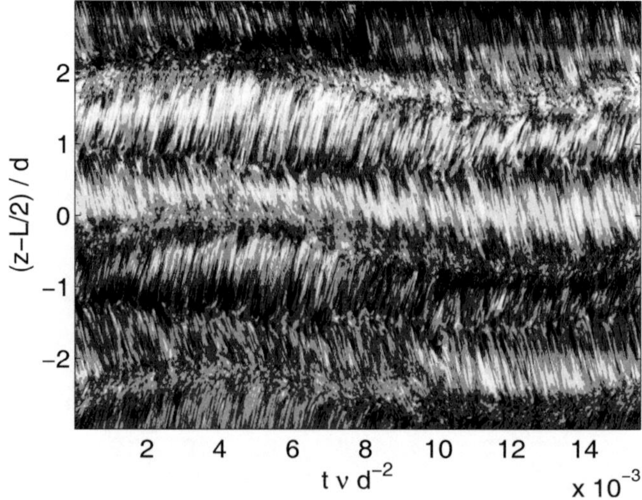

FIGURE 6.18: Space time diagram of turbulent Taylor-Couette flow for $Re_S = 52000$ and $\mu = -0.25..$ The diagram shows the first fragment of the measured data, the following data are given in Appendix A, Figures A.13, A.14, A.15.

As the systems rotation exceeds the counter rotation of the torque maximum ($\mu_{max} = -0.20$ see Chapter 4) the large-scale circulation will disappear with μ. In the closest investigated case, $\mu = -0.25$, the vortices are still strong 6.18. The already observed axial drift of the vortices becomes strong. Observing it over the whole time (Figures 6.18, A.13, A.14 and A.15) its obviously a long term wave. The amplitude of this shift is of the order of half a gap width. The waves period can be identified to be in the order of 10^{-2} viscous times ($T_{wave} \sim 10^{-2} t\nu d^{-2}$). The rotation rates of the cylinders ($n_1 = 2\pi\Omega_1 = 2.67Hz, n_2 = -0.67Hz$) are significant higher than this oscillation and not triggered by any asymetry.

Increasing the outer counter rotation to $\mu = -0.30$ the flow changes again. A large-scale circulation is still able to be identified but weakly (Figure 6.19). Over time a long term change is observable. In the beginning the axial downflow of the large-scale circulation is well suited while the upwashing is hard to observe in Figure 6.19, as time passes the downwashing becomes less observable (Figure A.16), and then a flow of axially localised turbulence can be seen in Figure Figures A.17 and at the end the flow changes to a well identified upwashing of the large-scale circulation while the downwash now is hardly observable (Figures A.18). This Scenario has even a longer wave length than the one of $\mu = -0.25$.

Finally the strongest counter rotation visualisation is performed is done for $\mu = -0.50$, Figure 6.20. The clear large-scale circulation is not observed. The flow is irregular, while bands of high illumination and low illumination are alternating along the axis.

6.1.4 Summary of flow patterns

The flow structures of counter rotating Taylor-Couette flow show an interesting behaviour. For the case of pure inner cylinder rotation one observes turbulent Taylor Vortices at low shear Reynolds numbers ($Re_S = 5000$). These Vortices remain also for slight counter rotation. As the outer cylinder rotation becomes more dominant the flow gets an axial dependency shifting the vortices up and down by an oscillation. The frequencies measured for these structures are much lower as the rotation rates of the cylinders but higher than the frequency of viscous time scale (μ_{Fluid}/d^2). The stabilizing effect of the outer cylinder rotation can be very well seen in Fig. 6.2 (left). This becomes important for the counter rotating Taylor-Couette flow. For certain cases the outer cylinders influence does not appear in the flow. At the point when the flow is turned and in this area the angular momentum distribution does suppress centrifugal instabilities (Rayleigh criterion) the angular motion transport gets limited. This results in a decrease of the torque. Hence, the idea of enhanced large scale circulation by

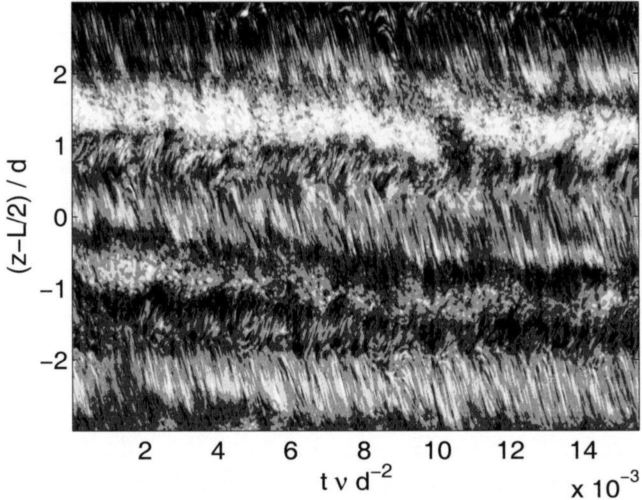

FIGURE 6.19: Space time diagram of turbulent Taylor-Couette flow for $Re_S = 52000$ and $\mu = -0.30$. The diagram shows the first fragment of the measured data, the following data are given in Appendix A, Figures A.16, A.17, A.18.

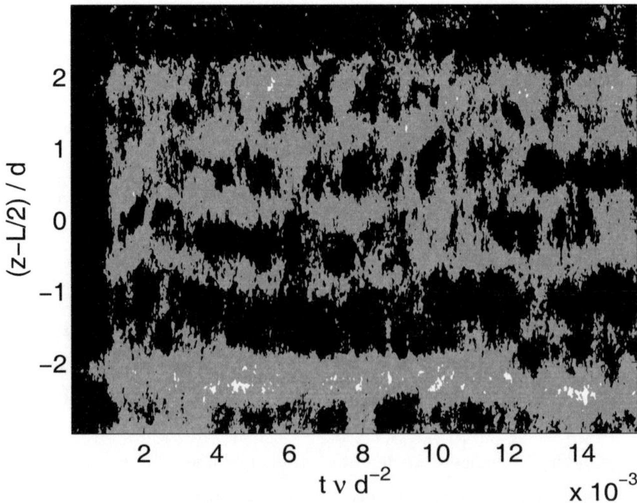

FIGURE 6.20: Space time diagram of turbulent Taylor-Couette flow for $Re_S = 52000$ and $\mu = -0.50$. The diagram shows the first fragment of the measured data, the following data are given in Appendix A, Figures A.19, A.20, A.21.

Brauckmann and Eckhardt [2] can be confirmed by our observations. The same holds for the larger shear Reynolds numbers. The main change by the increase of the driving, is the weakening of the large-scale circulation for pure inner cylinder rotation. For the case of slight counter rotation up to the maximal torque the large-scale circulation is enhanced. This indicates that the large-scale circulation carries the increase in angular momentum transport.

Chapter 7

Particle Image Velocimetry of near-wall azimuthal-axial flow

7.1 Introduction to near-wall Particle Image Velocimetry

To analyse the fluid flow, explained by the visualisation method in Chapter 6, in a quantified way, the videos of Kaliroscope particles observed have been processed using Particle Image Velocimetry. This leads to an understanding of the flow happening close to the wall. Using the Velocimetry it is possible to understand the large scale circulation as well as the turbulent fluctuations. The shear Reynolds number chosen for this investigation were $Re_S = 5000, 26000, 52000$ and 78000 changing the ratio of angular velocities in the values of $\mu = 0, -0.10, -0.15, -0.20, -0.25, -0.30, -0.50$.

7.2 Experimental procedure

Well known is the setup if a classical Particle Image Velocimetry [32]. A Laser illuminates particles in a certain plane in the fluid, while a camera captures the images of the particles. By the use of cross-correlation algorithm the displacement of the particles is processed and computed into velocities by knowing the time between the images were captured.

In the present case the aim is to analyse the fluid flow close to the outer wall and not inside the flow. As the outer wall is cylindrical, this leads to some difficulties by the use of a Laser sheet. Thus, the idea came to use an outer global illumination of the particles. To keep the sight at the outer wall, and no deep into the gap, a black dye is added to the fluid. By the concentration of the dye the transparency of the fluid can be adjusted.

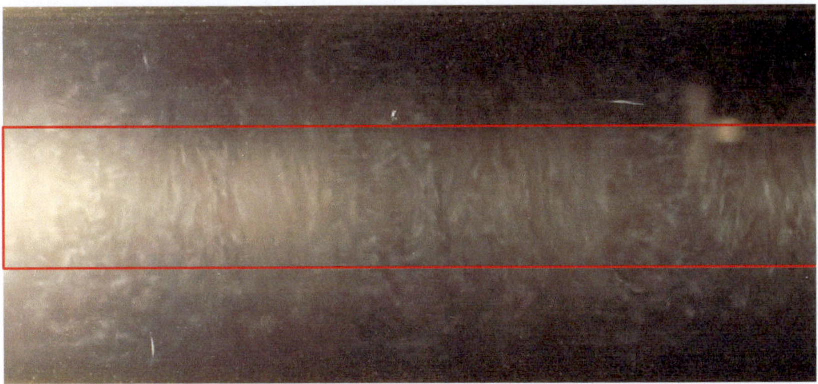

FIGURE 7.1: Image using visualisation technique with Kaliroscope particles and black dye. The red box indicates the area where the Particle Image Velocimetry has been processed. Camera used for this visualisation: Canon EOS 1100D (25fps, 1280x606 px used)

The concentration was chosen, to keep only a thin layer of about $2mm$ visible. Thus, the observable particles move with the polar coordinates with the cylindrical geometry close to the outer cylindrical wall. The fluid layer was observed by a Canon EOS 1100D camera capturing images with 25Hz or 30Hz. The frequency was adjusted depending on the flows observed. The cameras depth of focus was increased by closing the aperture to observe the particles in the cylindrical shape. Finally the particles were depicted in a good observable quality in an image size of about $15mm$ in azimuthal direction and $300mm$ in axial direction (respectively 0.4 gap widths in azimuthal and 8 gap widths in axial direction) for the case of $Re_s = 5000$. For the case of larger Reynolds numbers the field of view had to be narrowed to an observed axial length of $125mm$ leading to observable aspect ratio of 4.5 in axial direction and $7mm, (0.2d)$ in azimuthal direction. At this area the Particle Image Velocimetry was applied. It has to be noted that the field of view is relatively short in main velocity direction (azimuthal), but it enables an accurate measurement if the velocities close to the wall.

To calibrate pixel length into real scales, the calibration was done using a precise steel scale adjusted in axial direction of the experiment. To consider the cylindrical aberrations of the cylinder wall, the scale was traversed also in azimuthal direction. For the limit of $\pm 12mm$ around the perpendicular access of the system the aberrations are negligible. As already mentioned the maximum field of view was chosen to $\pm 7.5mm$ around the centre which is within this limit.

Using the fixed capturing rate of the camera, the number of samples for each case was in between 2000 and 3000 images. The particle shift between two following images was

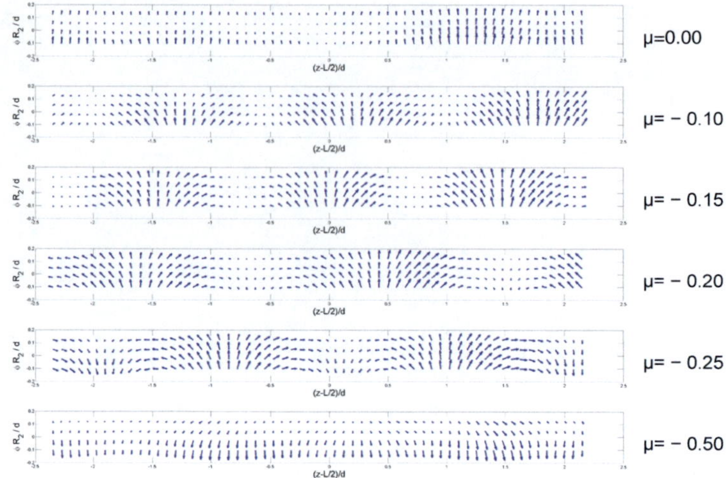

FIGURE 7.2: Mean velocity vector field of the near-wall PIV at $ReS = 26000$

processed using a refining PIV algorithm. The final interrogation areas were 16x16px with an overlap of 75%. The resulting mean velocity vector fields for shear Reynolds number $Re_S = 26000$ are depicted in Figure 7.2 for the different rotation ratios μ. In this representation it is quiet obvious that the large-scale circulation is strong for the case of counter rotation, as also observed by the visualisation (Chapter 6). For a better representation the both velocity components (angular velocity ω and axial velocity u_z) are shown in Figure 7.3 and Figure 7.5 while also the standard deviation of the angular velocity is given in Figure 7.4.

7.3 Results from Particle Image Velocimetry of near-wall azimuthal-axial flow

From these diagrams it is obvious, that for $\mu = 0$ turbulent Taylor vortices are weakly existing. The flow is mainly moving into the azimuthal direction due to the driving, with only a weak axial component and strong fluctuations. But especially in the temporal averaged axial component the large-scale circulation is observable.

To describe the mean flow the contour plots of the temporal averaged angular as well as axial velocity components are important, while the fluctuations 7.4 result in the turbulent kinetic energy in 7.6. Here the fluctuations only in azimuthal velocity is given but azimuthal component is similar in behaviour. For turbulent kinetic energy the local

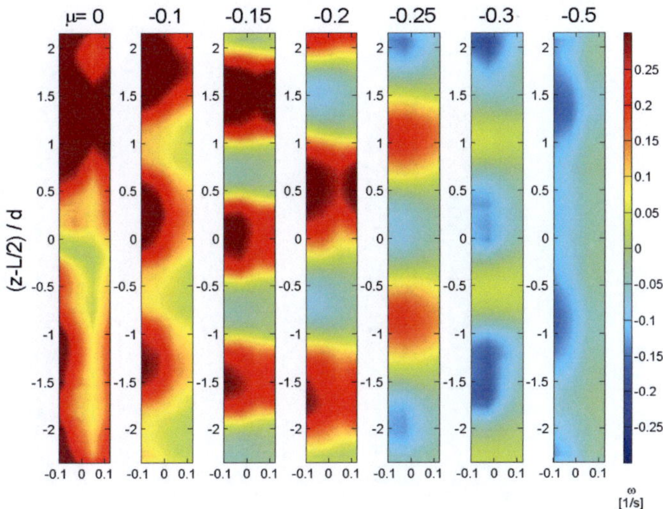

FIGURE 7.3: Contour plot of the temporal averaged angular velocity close to the wall
$ReS = 26000$

sum of square of both components is used for computation. As already mentioned, the radial component is not measured, while the size of radial velocity is negligible close to the outer wall.

For weak counter rotation ($\mu = -0.10$) already a strong effect in the flow is observable. Especially the axial velocity component due to the large-scale circulation is enhances. In The area of the outflow of the vortices also the turbulent fluctuations rise. As the large-scale circulation washes the fluid along the outer cylinder up- and downwards the flow becomes less turbulent up to the inflow region.

It is noteworthy, that, the temporal averaged angular velocity in the observed plane is moving with the direction of the inner cylinder. The outer cylinder is rotating in negative direction. Thus, the region of neutral velocity must be very close to the cylinder. As the counter rotation increases ($\mu = -0.15$) the axial component is again rising and the number of vortices does change in out campaign (nicely depicted in Figure 7.5). Again the turbulent fluctuations at the outflow region become stronger at the outer cylinder. At the inflow the flow is less turbulent. The mean flow still moves in angular velocity of the inner cylinder movement. The neutral surface, which must exist, can not be seen within the particles visible at the wall.

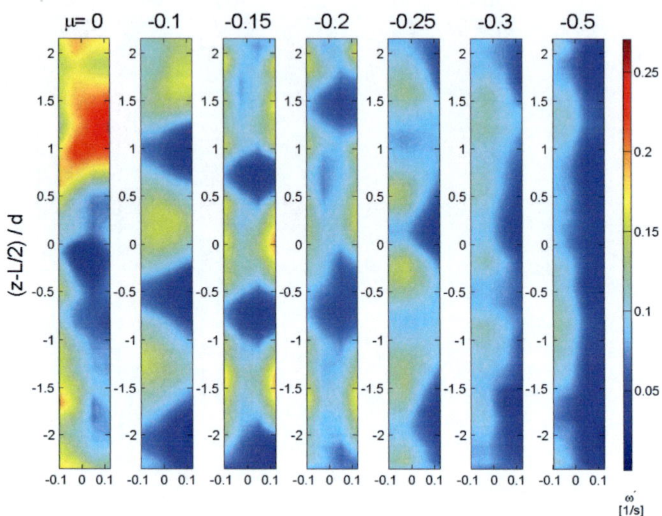

FIGURE 7.4: Contour plot of the standard deviation of the angular velocity close to the wall $ReS = 26000$

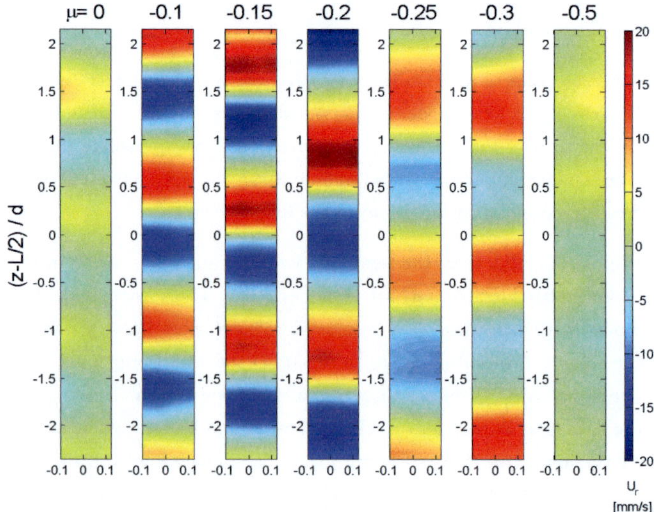

FIGURE 7.5: Contour plot of the average of the axial velocity close to the wall $ReS = 26000$

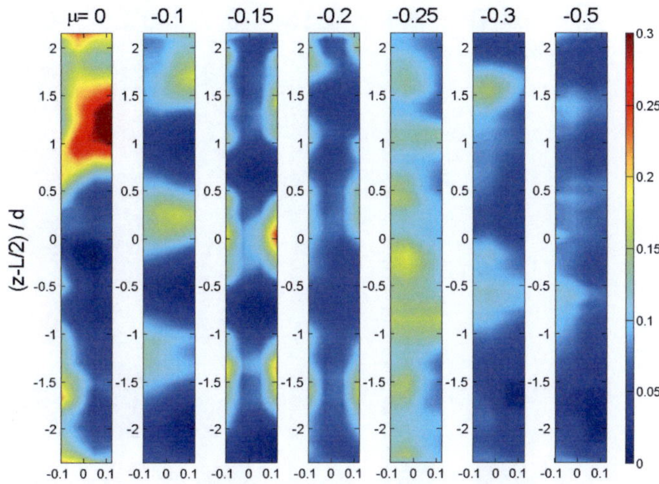

FIGURE 7.6: Contour plot of the turbulent kinetic energy of the local flow close to the wall $ReS = 26000$

At the Torque Maximum ($\mu = -0.20$) the large-scale circulation comes to its strongest enhancement. Again the cortices wave length did change, the vortices became bigger in axial direction. In Figure 7.6 one can see, that the fluctuations did reduce. At the inflow region now a counter flow is observed. But still in the region of the outflow the angular velocity is strongly positive. Even higher compared to the cases of weaker counter rotation. Whereas the angular velocity of the inner cylinder is reduced as the shear Reynolds number is constant and so, $\Delta\Omega = const.$.

As the counter rotation passes the maximum ($\mu = -0.25$) the flow changed drastically. Weaker axial velocity is observed in the Taylor vortices, the number of vortices rises. The turbulent fluctuations become stronger. The neutral phase now is moving away from the outer cylinder. But, at the outflow still the fluid contains high positive momentum (positive means in the direction of inner cylinder) reaches the outer cylinder wall. As it washes up/down with the large-scale circulation it changes its direction. Especially in the area of the inflow the fluid becomes very less turbulent. In this area a flow similar to a Couette flow can be imagined.

For the cases of even stronger counter rotation ($\mu = -0.30$ and -0.50) the sign of angular velocity changed to the one of the outer cylinder more and more, which is of course predictable. But, it also was predictable for the cases between $\mu = 0$ and $\mu = -0.20$ but did not happened so far. As the flow moves with outer cylinder rotation

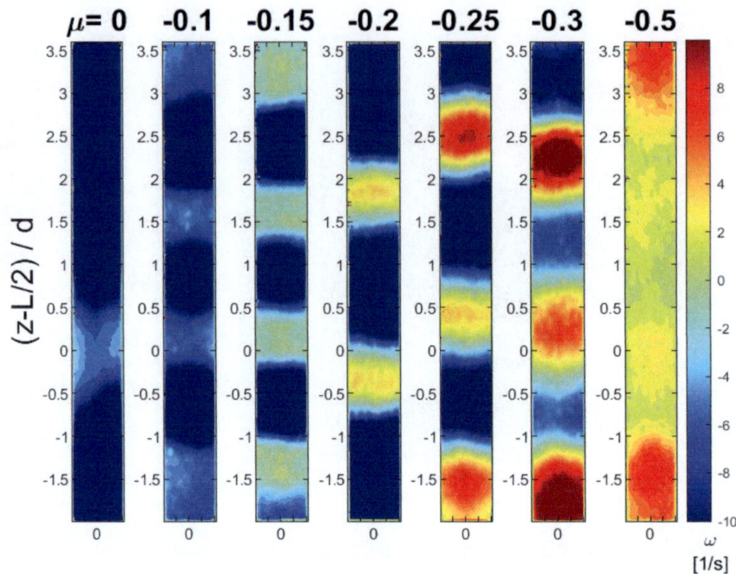

FIGURE 7.7: Contour plot of the temporal averaged angular velocity close to the wall
$ReS = 52000$

a low fluctuating velocity establishes and the large-scale circulation disappears at the outer wall. Thus, a Couette-like flow happens between the neutral phase and the outer cylinder. This area will suppress the angular momentum transport in radial direction. Thus, it reduces and the measured torques decrease.

7.3.1 Results for shear Reynolds number 52.000 and 78.000

The same procedure was also performed for the shear Reynolds numbers $Re_S = 5.2 \cdot 10^4$ and $Re_S = 7.8 \cdot 10^4$. Unfortunately the frame rate of the used camera was fix and not able to be adjusted on the higher velocities. Thus, the shear Reynolds number maximal observable kept at $7.8 \cdot 10^4$. The PIV algorithm could be still performed up to this shear Reynolds number. Although, the quality of the correlation values dropped significantly. Events, that the process could not compute sufficient velocities from the Particle images became more dominant. Thus, the errors from the correlation algorithm performed in PIV increased. Due to a huge number of images captured the statistical mean velocities can still be considered for analysis. While the standard deviation consists of the turbulent fluctuations as well as the errors. This leads us to only look to the mean

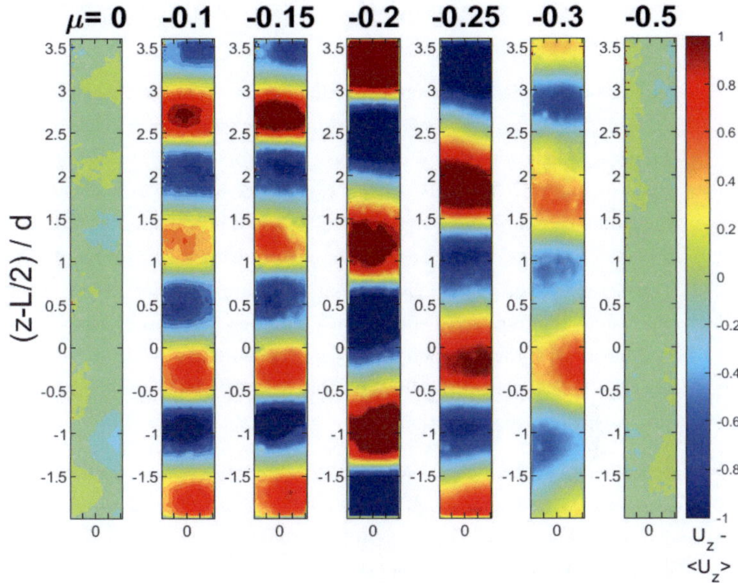

FIGURE 7.8: Contour plot of the average of the axial velocity close to the wall $ReS = 52000$

velocity components of the angular velocity as well as the axial velocity, see Figures 7.7, 7.8, 7.9, 7.10. To compute the turbulent kinetic energy from these measurements did not succeed as described.

For the shear Reynolds number $Re_S = 5.2 \cdot 10^4$ and outer cylinder at rest ($\mu = 0$) there are hardly no axial vortices identifiable in the angular and axial velocity component, see Figures 7.7, 7.8. Apart from a small acceleration at mid-height no dependency of the velocity was observed along the cylinder. As the counter rotation sets on ($\mu = -0.10$) immediately in the axial velocity component a periodicity can be seen. The axial velocity component is relatively strong and a wavelength of about $1.5d$ can be seen. Thus, the vortices are slightly contused. Following the mean angular velocity one can see, that in the areas of the outflow region (below positive axial velocity and above negative axial velocity), the high velocity from the inner part of the gap is advected onto the outer cylinder. The angular velocity decreases as it moves along the outer cylinder and at a smaller value it is advected by the large-scale circulation inward the system. Still in this case one can only observe angular velocity in the direction of the inner cylinder. As the ratio of angular velocities decreases more ($\mu = -0.15$) the axial velocity remains as in the case before. While the angular velocities are shifted down due to the lesser global rotation rate. The wave length kept the same as for $\mu = -0.10$, the advection of angular momentum becomes more visible.

For the rotation rate of the torque maximum $\mu = -0.20$ the behaviour does change (Fig. 7.7, 7.8). The axial velocity component increases significantly. The vortices become more dominant. Along with this also the wave length of the large-scale circulation increases to approximately $2d$. Also here a slight asymmetry of the vortices against the mid-height can be observed as already seen in Chapter 5.2. Thus, the results of both campaigns are of good agreement. Again angular momentum with the direction of the inner cylinder is advected throughout the gap and hits the outer cylinder. There it travels along the axis due to the large-scale circulation - here now the sign of velocity changes and the fluid moves with the direction of the outer cylinder in a relatively short axial domain of less than half a gap width. In Total this is again the case with the largest expanse of each vortices in axial direction.

As the outer cylinder rotation becomes more dominant ($\mu = -0.25$) the axial velocity decreases in its magnitude (Fig. 7.7, 7.8). Also the axial wavelength again changes slightly. While still the wavelength is of about $2d$, the positions of the central upward and downward component changed. As the counter rotation becomes more dominant ($\mu = -0.30$), the axial wave remains but the axial velocity component decreases more. In the case of $\mu = -0.50$ the axial velocity behaves similar to the one of $\mu = 0$, there is no large scale circulation seen. Along that also the angular velocity shifts towards the rotation of the outer cylinder. As the rotation of the outer cylinder becomes more dominant for the fluid close to the outer wall also the differences in the angular velocity between an outflow and inflow region decrease. Finally at $\mu = -0.5$ the angular velocity seems to be relatively uniform.

For the highest studied shear Reynolds number of $Re_S = 78.000$ it is noteworthy, that this shear Reynolds number is considered as the region where the change between the classical turbulent Taylor-Couette flow (where the bulk is turbulent but the boundary layers remain laminar) and the ultimate turbulent Taylor-Couette flow (where bulk flow as well as boundary layers are turbulent) takes place. The determination of this was done using the torque data as the change in the exponential behaviour can be identified in this region. While unfortunately no torque data was available in this area. The reason for that had pure technical background as for given viscosity of the fluid the range of measurable torque is limited between the driving velocity where the torque meter senses enough torque to be qualified and the upper limit where no faster rotation was possible.

A more spatial dependence of the angular and azimuthal velocity components can be nicely identified from the PIV of near-wall azimuthal-axial flow, depicted in Figures 7.9, 7.10. Again for pure inner cylinder rotation, $\mu = 0$, no axial wave can be observed. But,

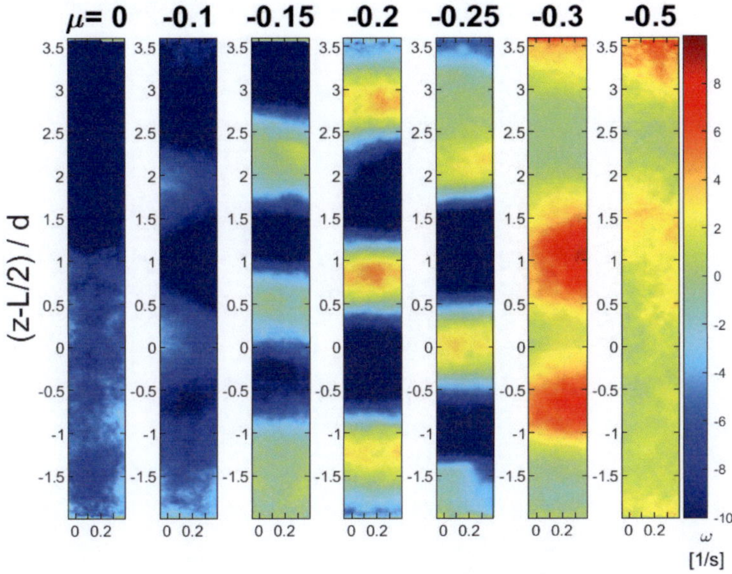

FIGURE 7.9: Contour plot of the temporal averaged angular velocity close to the wall
$ReS = 78000$

the axial velocity is slightly tending upwards above mid-height and directed downwards below mid-height. A clear large-scale circulation is not given.

As the outer cylinder rotation sets on and moved to the opposing direction the same feature observed in the cases before can also be seen for this large shear Reynolds number (Figures 7.9, 7.10). A wavelength of about 1.5 gap widths can be extracted from the flow field. The axial velocity set in is dominant and advects angular velocity along the outer cylinder wall. Also in this case the observed material moves with the direction of the inner cylinder and is dominated by the instability of the Rayleigh-criterion. As the rotation ratio decreases to $\mu = -0.15$ the positions of the turbulent Taylor-vortices change slightly. The wave length increases to about 1.7 gap widths, the strength of the axial velocity stays similar to the case before. The angular velocity is advected with the large-scale circulation from the inner cylinder and fast material meets the outer cylinder. Along the axial movement the deceleration takes place. Again in this case the flow even at the inflow region, can be identified as in the direction of the inner cylinder. The neutral phase, the position where the velocity turns sign, must be within a very thin layer at the outer cylinder. This layer must be less than $0.5mm$ as the fluid of a layer of about $2mm$ is observed in this investigation.

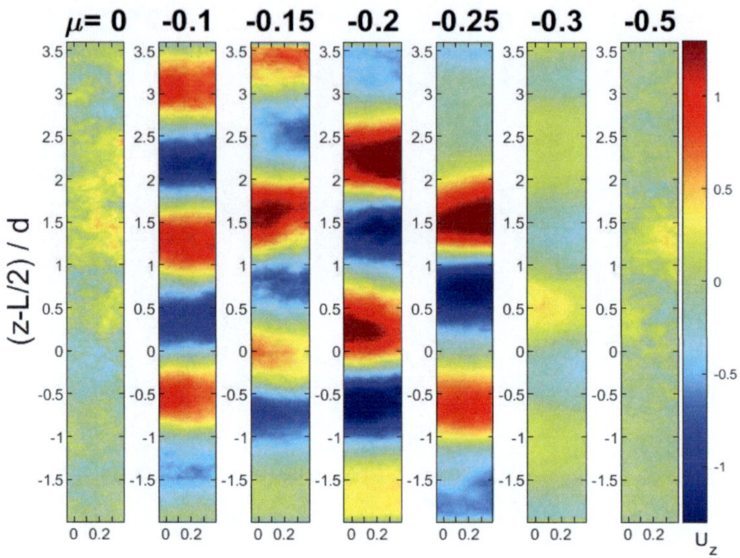

FIGURE 7.10: Contour plot of the average of the axial velocity close to the wall $ReS = 78000$

For the case of torque maximum, $\mu = -0.20$, the axial flow of the large-scale circulation increases Figures 7.9, 7.10). The wave length is changed again to 2 gap widths. Thus, the vortices are relatively symmetrical in axial as in radial direction. Here, the contrast in the advected high-momentum fluid from the inner cylinder, getting to the outer cylinder, towards the flow advected inward is the strongest. At the inflow flow changed the direction and the neutral phase detached from the outer cylinder. These vortices advect a lot of momentum. Due to the advection the profiles are flattened strongly and the steepness of the gradient at the wall must increase. This leads to the larger drag at the wall, which is the cause of the increase in the measured torques.

Beyond the torque maximum the large-scale circulation becomes irrelevant (Figures 7.9, 7.10). For $\mu = -0.25$ the axial waves again change. It even seems that the direction of the large-scale circulation did change. The variation in the axial velocity component between up- and down-flow decrease and the advection of velocity as well. For $\mu = -0.30$ the identification of the large-scale circulation becomes less significant in both velocity components and drop entirely for the rotation ratio of $\mu = -0.50$.

7.3.2 Energy contributions to the transport process

One can also compute from the local turbulent fluctuations a turbulent kinetic energy averaged over the observed segment of a cylinder (E_{Tur}). Thus, an integral is computed on the local fluctuations. For each ratio of angular velocities this is depicted in Figure 7.11 in green. The energy in the large scale circulation at the outer wall is computed by $E_{LSC} = < \bar{u}_z^2 >_{\phi,z,t}$. This and the sum of both ($E_{Tur} + E_{LSC}$) are also depicted in Figure 7.11 as blue and red respectively. It has to be mentioned, that this analysis is only valid for this shear Reynolds number. The other cases are affected by the errors, which are also given in the statistical value of the local turbulent fluctuations. Thus, the other shear Reynolds numbers the PIV was performed are excluded.

The behaviour of the total energy is very similar to the one of the measured torque. It has to be concluded: for pure inner cylinder rotation for turbulent Taylor-Couette flow the large-scale circulation is very weak compared to the turbulent fluctuations. As the counter rotation establishes, on one hand the large-scale circulation enhances and on the other hand the turbulent fluctuations drop. As the fluctuations stay relatively of the same order with stronger counter rotation, the Energy in the large-scale circulation increases up to the value of $\mu = -0.2$, which is the torques maximum. Interestingly it decreases strongly from $\mu = -0.20$ to $\mu = -0.25$, while in this case the turbulent fluctuations do rise again. For even stronger counter rotation finally the Turbulent kinetic energy remains at a similar value but the Energy in the large-scale circulation decreases significantly.

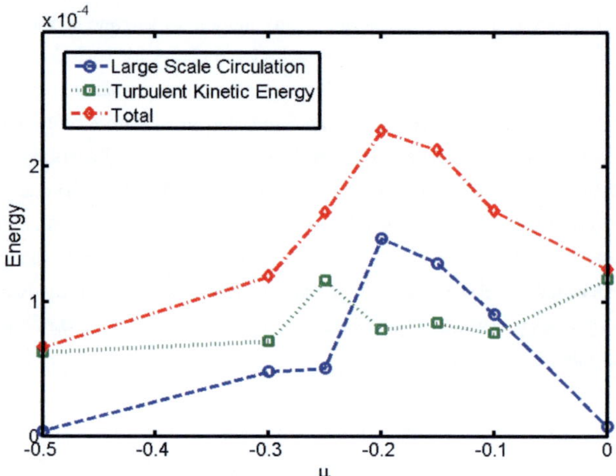

FIGURE 7.11: Energy contributions processed using the PIV from flow visualisation close to the wall $ReS = 26000$

Chapter 8

Particle Image Velocimetry of Azimuthal-Radial flow

8.1 Experimental setup of the PIV in azimuthal-radial planes

In this Chapter measurements by means of Particle Image Velocimetry in azimuthal planes are used to derive velocity profiles, flow structures and scaling in the wide-gap turbulent Taylor-Couette flow. The measurements are carried out in the Top-view Taylor-Couette Cottbus experiment for various flow conditions, depicted in Figure 8.3 left. For these measurements the transparent top plate was used and the camera installed above the experiment looking into the cylinder gap in axial direction. The LASER sheet is placed through the cylinder and illuminates an azimuthal-radial plane, one example of the resulting velocity field is given in Figure 8.4. The LASER is placed on a traverse giving the ability of changing the height of the LASER and changing the position of the analysed measurement plane. The camera is refocused to each plane. For the measurements two different resolutions of the axial spacing are chosen, based on the investigations done before. The axial length from top to bottom plane is given to be $80mm$. As the gap width is $d = 35mm$ the length observed is basically larger than two gap widths. The wavelength of the vortices was mostly in the size of $2d$, so in all measurements at least one pair of Taylor-Vortices would be expected in the measured plane. Due to the axial traversing on each height the displayed cylindrical gap changes in terms of the pixel dimensions of the sensor. Thus, for each height a calibration had to be performed. The resulting scale factor for the different heights is given in Figure 8.3 right. Due to this, the measurements were limited in length resulting in the maximal path of $80mm$. The measurement parameter space is depicted in Figure 8.3 left for two different used spacings of the measurement planes. A huge number of different shear

Reynolds numbers $4 \cdot 10^4 \leq Re_S \leq 2.7 \cdot 10^4$ has been investigated for outer cylinder at rest ($\mu = 0$). From Chapter 7 it is already known, that for the case of pure inner cylinder rotation the large-scale circulation is weakly given. Thus, for these cases the spacing is set to be $8mm$. Another set of data is done for the fix shear Reynolds numbers $Re_S = 8 \cdot 10^4$ and 10^5, while the rotation ratio is varied from $0.2 \geq \mu \geq -0.6$ with an axial spacing of the measurement plane of $4mm$. In addition for the case of the rotation ratio the torque is maximal ($\mu = -0.2$) also for larger shear Reynolds numbers measurements with this spacing were carried out.

FIGURE 8.1: An example of the particle images recorded. The inner cylinder is visible at the left side, while the outer cylinder is seen at the right hand side. The picture is taken at $Re_S = 100000, \mu = -0.40$.

To perform the Particle Image Velocimetry a double pulsed Nd:YAG LASER with wave length of $532nm$ and an energy of $100mJ$ per pulse is used to generate a light sheet of $2mm$ thickness. The time between pulses is arranged in the way that for each measurement sufficient displacement of about 4 to $10px$ is observed, resulting in intervals of $70\mu s$ up to $870\mu s$. The main frequency of the system is fixed to $15Hz$, which is the LASER's maximum working frequency. Thus, the measurements are not time resolved. The used camera is a Flow Sense EO2M with a CCD resolution of 1200×1600 pixels. Each measurement consists of 150 double images for one height, resulting in 10 seconds measurement time. The limitation is done with respect of the huge number of heights analysed while the flow properties should remain stable.

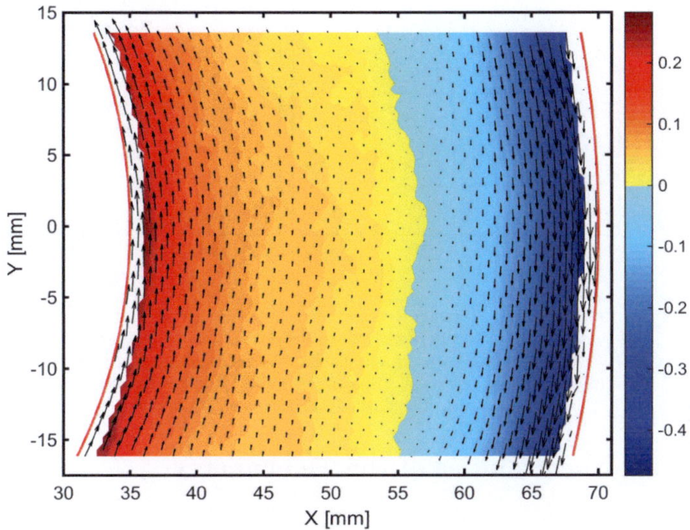

FIGURE 8.2: An example of the Particle Image Velocimetry. The inner cylinder is visible at the left side, while the outer cylinder is seen at the right hand side. The picture from Figure 8.1 and the followed double image is used to process this vector field at $Re_S = 100000, \mu = -0.40$. Here only every fourth vector is displayed.

As working fluid distilled water is used, with a kinematic viscosity of $\nu(20) = 10^{-6}m^2s^{-1}$. Hollow glass microspheres, with a size of $10 - 20\mu m$, are used as tracer particles. The resulting Stokes number for these particles is below $4 \cdot 10^{-3}$, giving them the ability to follow the flow. The fluid temperature does increase due to friction and is measured before and after each run including the 21 height steps, the fluid temperature change did not exceed $1K$. Thus, the temperature can be considered to be stable.

The double frame particle images are captured (Figure 8.1) and analysed using Dantec Dynamics Studio v.3.31. An adaptive PIV algorithm is performed and optimizes the size and shape of each interrogation area (IA) depending on local flow gradients and seeding densities. The approach of using multiple grid resolutions results is a final vector spacing of 16px for each height. Thus, the spatial resolution is depending on the measured height with the computed scale factor in Figure 8.3 right. Figure 8.2 shows an example of one processed snapshot, the temporal average is given in Figure 8.4.

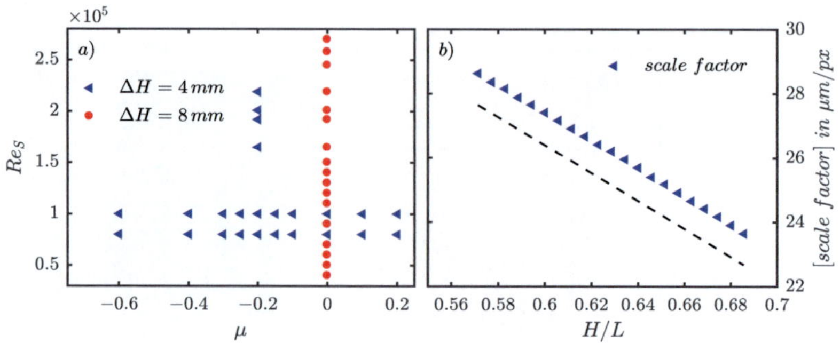

FIGURE 8.3: Left: Parameter space investigated using the PIV in azimuthal plane at different heights in terms of the shear Reynolds number Re_S and the rotation ratio μ. Blue triangles represent measurements over an axial length of $80mm$ by separation of $4mm$; red circles represent a spacing of $8mm$. Right: Scale factor at the 21 height positions. The dashed line serves as a guide for the eye.

8.2 Uncertainties of the Velocimetry

As in every measurement also the uncertainties in Particle Image Velocimetry have to be considered. The used PIV algorithm uses a subpixel interpolation scheme for the detection of the correlation peak. Following Nobach and Bodenschatz [26] the resolvable displacement for each velocity computation is limited to $\Delta_{SM} = 0.1px$. In terms of the statistics, the standard error of the average Δ_{mean} reduces due to the number of samples. For shear Reynolds number $Re_S = 10^5$ the shear velocity is $u_s = u_2 - u_1 = 2.85ms^{-1}$, where $u_{1,2}$ are the azimuthal velocities of inner and outer cylinder. The resulting uncertainties lie exemplarily for:

$\mu = +0.20 : \Delta_{SM} < 24mms^{-1}, \Delta_{meant,\varphi} < 0.23mms^{-1}$, for

$\mu = 0.00 : \Delta_{SM} < 10mms^{-1}, \Delta_{meant,\varphi} < 0.1mms^{-1}$, for

$\mu = -0.20 : \Delta_{SM} < 5mms^{-1}, \Delta_{meant,\varphi} < 0.04mms^{-1}$ and for

$\mu = -0.40 : \Delta_{SM} < 3.6mms^{-1}, \Delta_{meant,\varphi} < 0.03mms^{-1}$.

Thus, the error of a single measurement does not exceed 0.9% of the shear velocity and the standard error of the temporal and azimuthal averaged velocity is below 0.01%.

Another source of uncertainties comes due to the use of only 150 snapshots per measurement position, which is relatively short in time. This had to be chosen as for one flow parameter 21 heights were analysed and the flow properties had to be kept constant. Nevertheless, for turbulent flows a larger number of snapshots would be vital for statistically converged results. As we can compute the quasi Nusselt number also from the velocity fields (see Section 8.8) it gives us ability to quantify the convergence. This has been done for the highest shear Reynolds numbers for outer cylinder at rest as well

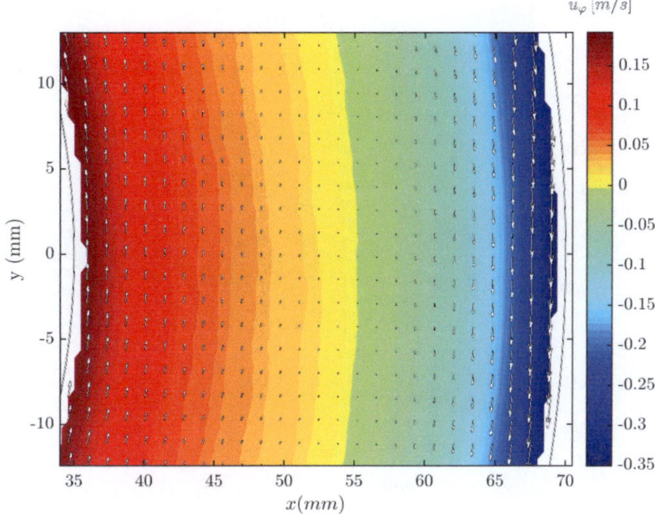

FIGURE 8.4: Temporal and axial averaged velocity field for shear Reynolds number $Re_S = 100000$ and ratio of angular velocities is $\mu = -0.40$. At 21 different heights a PIV plane was investigated. For better representation only every fourth vector in both directions is displayed.

as the rotation ratio at torque maximum. After 8 seconds of measurement time the Nusselt number is already converged. Taking the remaining two seconds into account to compute the standard deviation related to its mean, the statistical uncertainty can be quantified to be 0.24% and 0.12% for the mentioned cases of largest shear Reynolds number. Considering, that the flow had to be stable from start of top measurement until the end of bottom measurement the chosen number of snapshots is large enough for reliable results.

8.3 Flow fields and velocity profiles

The given flow fields reveal numerous interesting structures of the fully three dimensional flow. As the heights could not be measured simultaneously the temporal statistical flow has to be considered for the analysis of the three dimensional flow. Figure 8.5 shows temporal averaged radial velocity component in an overlay of all 21 measured planes for the shear Reynolds number of $Re_S = 8 \cdot 10^4$ and rotation ratio of $\mu = -0.20$, corresponding to the torque maximum. Obviously outflow as well as inflow regions of the large-scale circulation can be identified. The plane in x-z coordinates finally shows

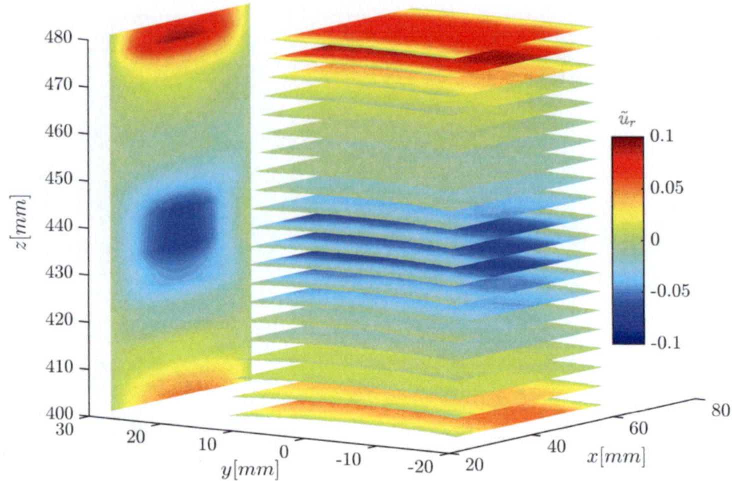

FIGURE 8.5: Temporal averaged radial velocity in different heights analysed for shear Reynolds number $Re_S = 80000$ and ratio of angular velocities is $\mu = -0.20$. At 21 different heights a PIV plane was investigated. The resulting meridional plane is depicted from the temporal, azimuthal average of each plane investigated.

the contour plot of the temporal and azimuthal averaged radial velocity component, respectively in the radial-axial plane.

In the following the radial coordinate is normalized by the gap width as $r - R_1/d$, leading to 0 at inner cylinder wall and 1 at the outer cylinder wall. The angular velocity is plotted in terms of the difference of the systems rotation $\Delta\Omega = |\Omega_2 - \Omega_1|$ in that way, that the outer cylinders velocity represents $\omega(2) - \Omega_2/\Delta\Omega = 0$ and one for the inner cylinders angular velocity $(\omega(1) - \Omega_2/\Delta\Omega = 1$. Based on the shear Reynolds number a shear velocity is defined $(u_s = Re_S\nu d^{-1})$ to normalize the radial velocity component.

Following the velocity profiles of the temporal, azimuthal averaged angular velocities and radial velocities at different heights are analysed in more detail (see for example Figure 8.6). Also the turbulent fluctuations are computed by the standard deviation of the radial velocity in time and azimuthal direction (see exemplarily Figure 8.7. The reconstruction of the meridional flow depicted in Figure 8.5 is also done for all flow parameters (see exemplarily Figure 8.8).

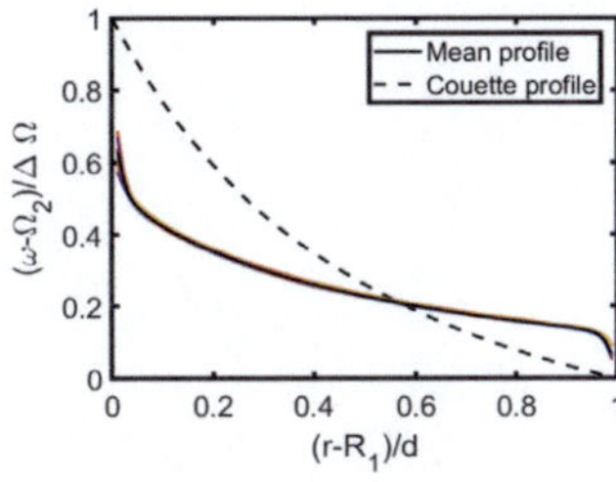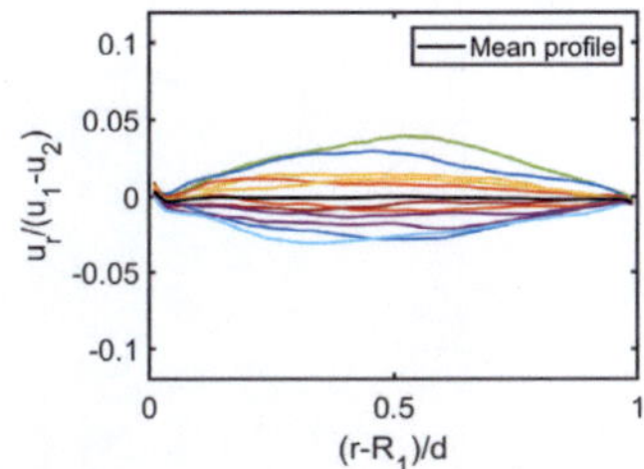

FIGURE 8.6: Angular and radial velocity profiles averaged in time and azimuthal coordinate for $Re_S = 40000$, $\mu = 0$. In colour the different axial positions are depicted, the black solid line represents the axial mean profile and the dashed line in angular velocity profile represents the laminar Couette-solution.

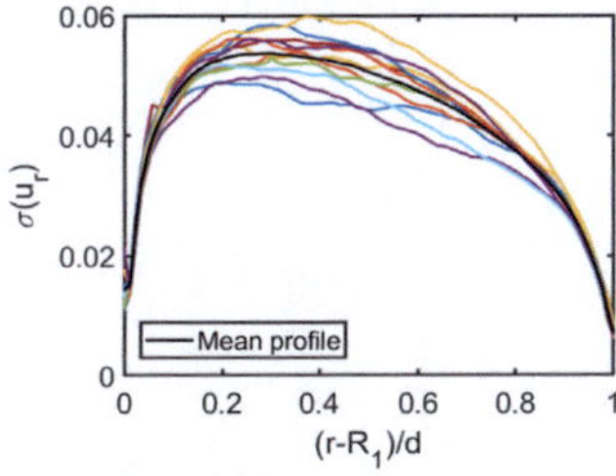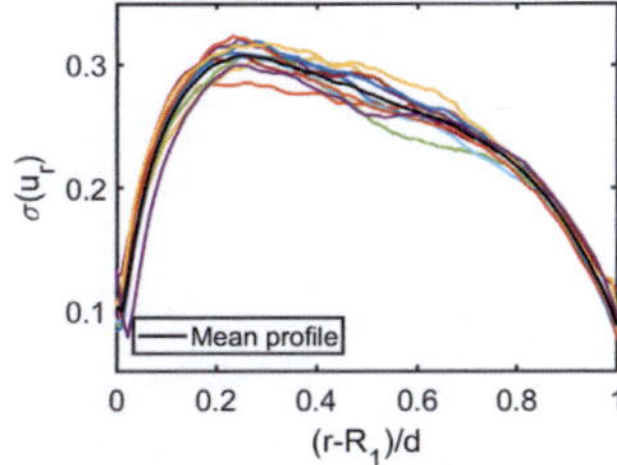

FIGURE 8.7: Profiles of the standard deviation of the radial velocity component computed in time and azimuthal coordinate for $Re_S = 40000$ (left) and $Re_S = 270000$ (right) $\mu = 0$. In colour the different axial positions are depicted, the black solid line represents the axial mean profile.

8.4 Influence of the shear Reynolds number for pure inner cylinder rotation

For the case of keeping the outer cylinder at rest a large number of different shear Reynolds numbers has been investigated. The torque was already discussed in Chapter 4 for the case of $\mu = 0$, where also a huge number of shear Reynolds numbers is used for analysing the scaling of the torque with the shear Reynolds number. Here now the flow behaviour is discussed in more detail. The resulting velocity profiles for $Re_S = 4 \cdot 10^4$ are depicted in Figures 8.6. For the measured heights the angular velocity profiles are very similar and hardly to be seen beyond the mean over all heights. The axial velocity component does vary for the different heights. The reconstruction onto a meridional plane is shown in 8.8, here it can be seen, that only a weak secondary flow is observed. The standard deviation profile of this flow is depicted in 8.7 left. For increasing shear Reynolds number the behaviour of the radial velocity component remains the same

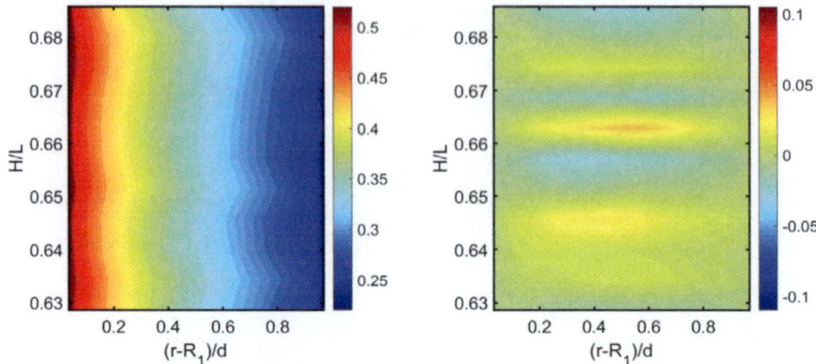

FIGURE 8.8: Contours of the Angular (left) and radial (right) velocities averaged in time and azimuthal coordinate for $Re_S = 40000$, $\mu = 0$.

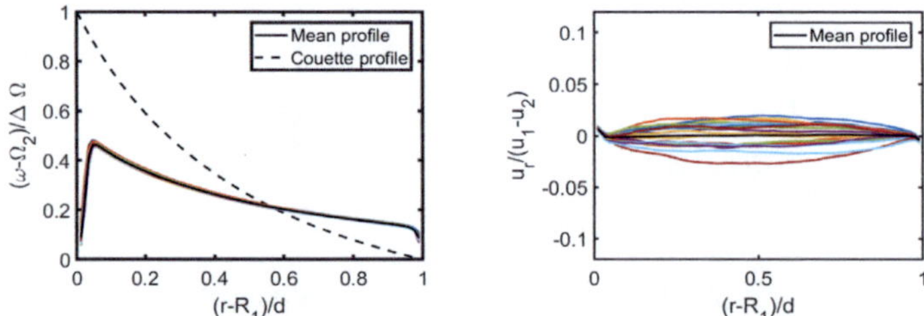

FIGURE 8.9: Angular and radial velocity profiles averaged in time and azimuthal coordinate for $Re_S = 80000$, $\mu = 0$. In colour the different axial positions are depicted, the black solid line represents the axial mean profile and the dashed line in angular velocity profile represents the laminar Couette-solution.

until shear Reynolds number of $Re_S = 8 \cdot 10^4$ (Figure 8.9). Passing this shear Reynolds number causes a drastic change in the radial velocity component. Figure 8.10 shows that already for $Re_S = 9 \cdot 10^4$ the radial velocities completely broke down. This remains for all following Re_S up to the maximum measured one, see Figures 8.11 and 8.12. For various shear Reynolds numbers the measurement is performed and all profiles of average velocities as well as the standard deviation are given in Appendix B Figures B.1 until B.19 for completeness.

Also the standard deviation of the radial velocity component, which can be identified with the wind Reynolds number Re_w (compare Chapter 5), has a very similar behaviour along the analysed Re_S. Between $Re_S = 4 \cdot 10^4$ (Fig. 8.7 left) and $8 \cdot 10^4$ a strong axial variation can be observed (Fig. 8.13 left). Also the radial dependency is strongly curved along the radii, especially in the centre of the flow. From $Re_S = 9 \cdot 10^4$ (Fig. 8.13 right) up to the maximum of $2.7 \cdot 10^5$ (Fig. 8.7 right) the axial dependency becomes weaker

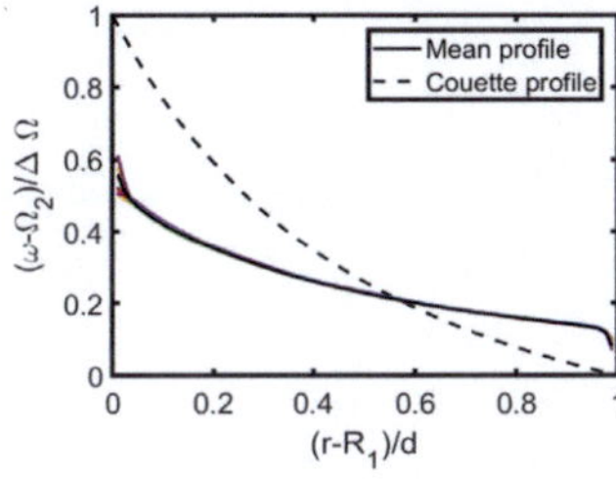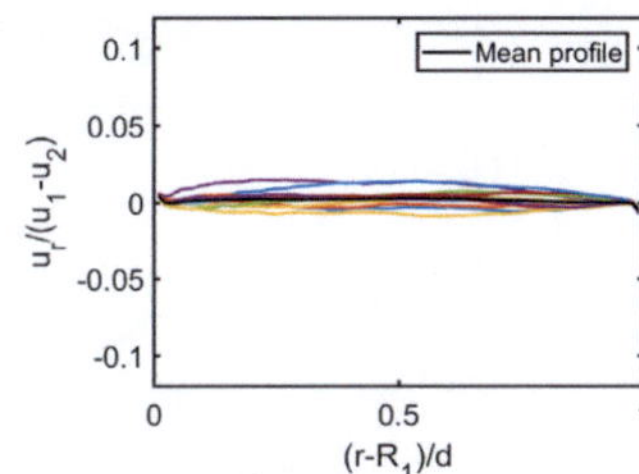

FIGURE 8.10: Angular and radial velocity profiles averaged in time and azimuthal coordinate for $Re_S = 90000$, $\mu = 0$. In colour the different axial positions are depicted, the black solid line represents the axial mean profile and the dashed line in angular velocity profile represents the laminar Couette-solution.

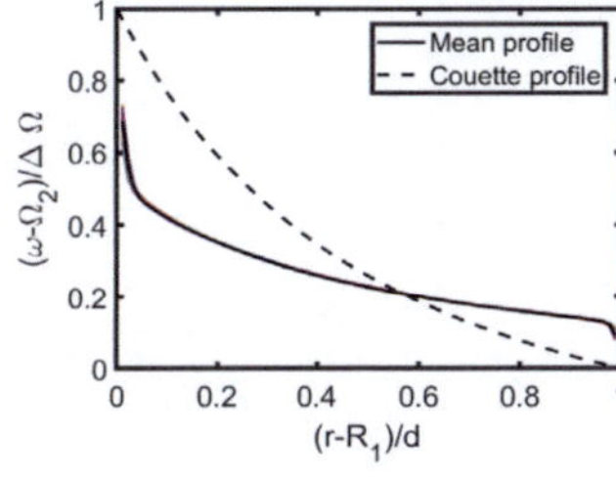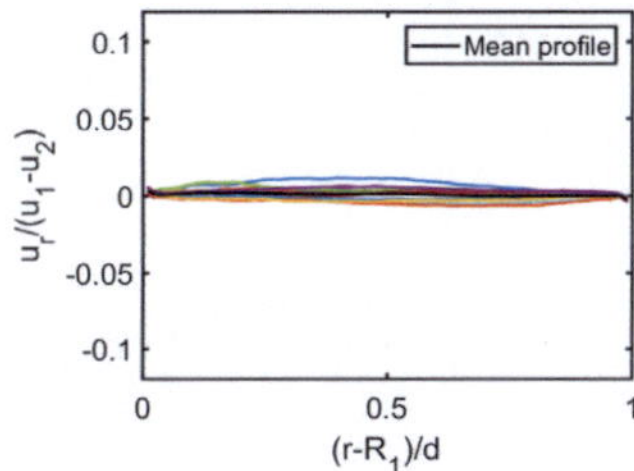

FIGURE 8.11: Angular and radial velocity profiles averaged in time and azimuthal coordinate for $Re_S = 150000$, $\mu = 0$. In colour the different axial positions are depicted, the black solid line represents the axial mean profile and the dashed line in angular velocity profile represents the laminar Couette-solution.

while in the central region of the radial profile the curvature of the dependency decreases. Also in this case the complete set of profiles is depicted in Appendix B Figures B.46 until B.54. A convolutee of the temporal, azimuthal and axial averaged velocity profiles for the given shear Reynolds numbers are depicted in Figure 8.15.

Thus, in the case of the pure inner cylinder rotation, the large-scale circulation disappears with increasing shear Reynolds number. A weak radial flow can still be observed up to $Re_S = 8 \cdot 10^4$. Above this the radial flow is to weak to be identified. Interestingly at the same region of shear Reynolds number, the scaling in the torque changes as well. This was considered as the change between the classical turbulent regime, where bulk flow is turbulent but the boundary layers not, to the ultimate regime, where bulk flow as well as boundary layers are turbulent. In this measurements the boundary layers were not resolved sufficient to judge about the turbulent fluctuations inside. But it can be seen, that the bulk flow changes when the scaling of the torque changes its exponent.

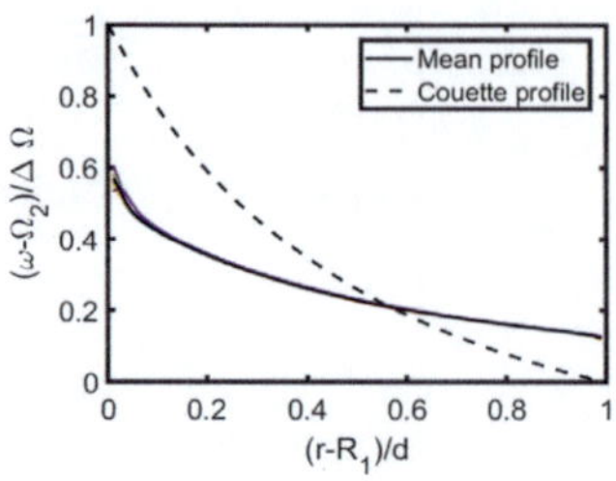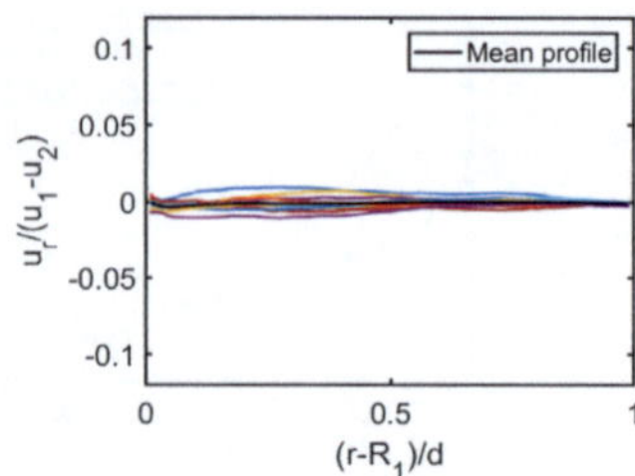

FIGURE 8.12: Angular and radial velocity profiles averaged in time and azimuthal coordinate for $Re_S = 270000$, $\mu = 0$. In colour the different axial positions are depicted, the black solid line represents the axial mean profile and the dashed line in angular velocity profile represents the laminar Couette-solution.

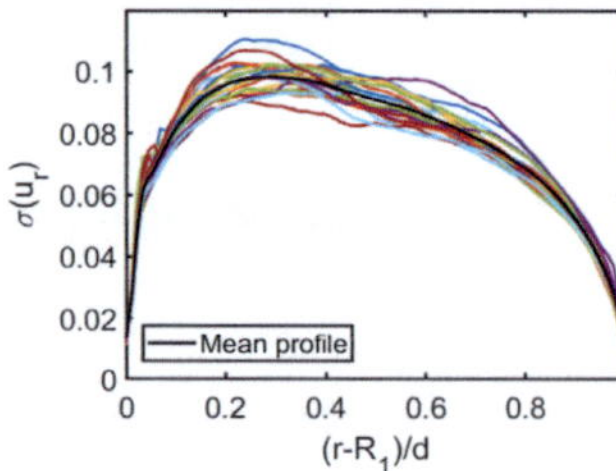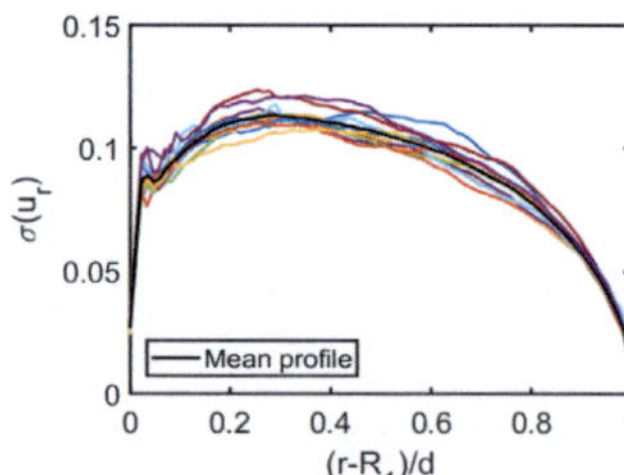

FIGURE 8.13: Profiles of the standard deviation of the radial velocity component computed in time and azimuthal coordinate for $Re_S = 80000$ (left) and $Re_S = 90000$ (right) $\mu = 0$. In colour the different axial positions are depicted, the black solid line represents the axial mean profile.

As the axial variation is not that big it is hardly no effect to be seen in the contour plots of the angular velocity and radial velocity component depicted as contour plots in Figure 8.8 and 8.14. This will be different for the flows, where the rotation ratio μ differs from 0.

8.5 Influence of the shear Reynolds number at the torque maximum

For counter rotating cylinders the turbulent Taylor-Couette flow changes drastically. The torque is maximized following Chapter 4 at a rotation ratio of $\mu_{max} = -0.20$ for the radius ratio $\eta = 0.5$, studied here. In contrast to the pure inner cylinder rotation ($\mu = 0$) the velocity profiles also reveal a change in the flow (see Figure 8.16). Obviously a very strong radial flow is identified resulting in a drastically change of the angular velocity profile as well. Due to the stronger radial mean flow the angular momentum is transported more efficient between different radii. Thus, the angular velocity profile

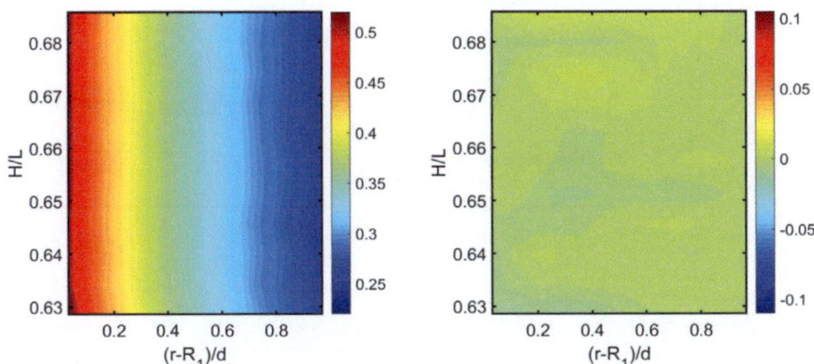

FIGURE 8.14: Contours of the Angular (left) and radial (right) velocities averaged in time and azimuthal coordinate for $Re_S = 270000$, $\mu = 0$.

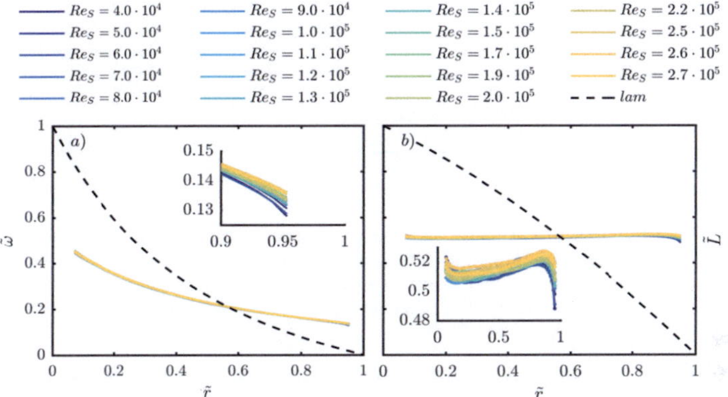

FIGURE 8.15: Angular velocity and angular momentum profiles along the radius of the temporal, azimuthal and axial averaged velocity fields for shear Reynolds numbers $Re_S = 4 \cdot 10^4$ until $2.7 \cdot 10^5$. The angular velocity is normalized as $\tilde{\omega} = (\omega - \Omega_2)/\Delta\Omega$ and the angular momentum respectively $\tilde{L} = (L - L_2)/(L_1 - L_2)$. The dashed line represents the laminar Couette solution.

becomes flattened. As the shear Reynolds number increases the behaviour does not change (Figure 8.17). The velocity profiles are depicted for completeness in Appendix B Figures B.31 to B.25.

Plotting the profile of the deviation of radial velocity in Figure 8.18 for $\mu = -0.20$ a big difference to $\mu = 0$ can be seen here as well. In the bulk the system has a relatively flat behaviour while it increases strongly close to the walls. A peak at each wall can be distinguished. It is again noteworthy, that the measurement can not claim to resolve the inner and outer boundary layer. Thus, these peaks are not to be understood as the inner

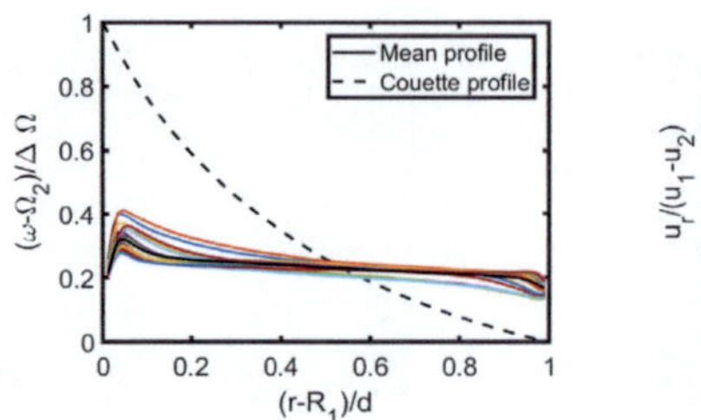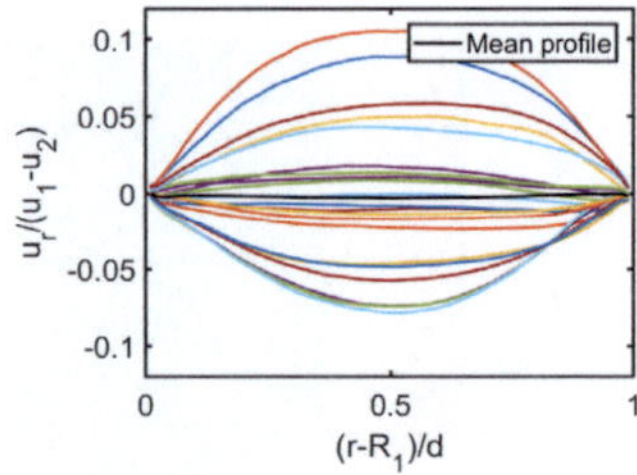

FIGURE 8.16: Angular and radial velocity profiles averaged in time and azimuthal coordinate for $Re_S = 80000$, $\mu = -0.20$. In colour the different axial positions are depicted, the black solid line represents the axial mean profile and the dashed line in angular velocity profile represents the laminar Couette-solution.

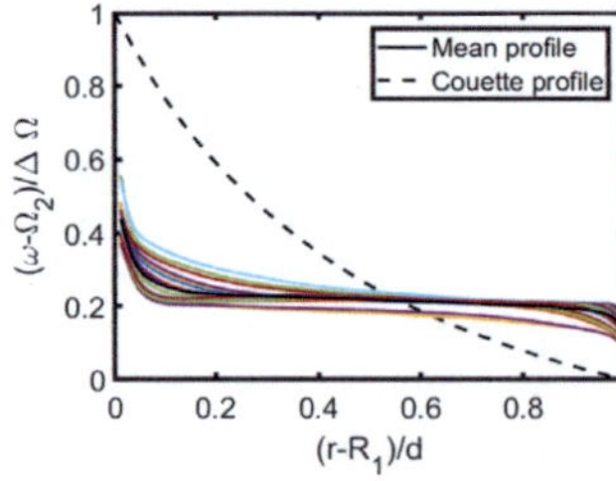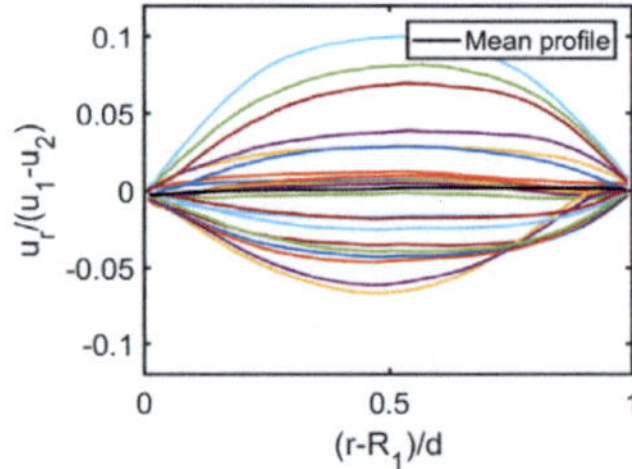

FIGURE 8.17: Angular and radial velocity profiles averaged in time and azimuthal coordinate for $Re_S = 201000$, $\mu = -0.20$. In colour the different axial positions are depicted, the black solid line represents the axial mean profile and the dashed line in angular velocity profile represents the laminar Couette-solution.

turbulent peak of the boundary layer. Especially as this peak becomes more visible as the shear Reynolds number increases and the boundary layer thickness should decrease, see also Appendix B Figures B.55 - B.57. Also a stronger dependence on axial positions can be seen. The deviations do deviate in axial direction, as observed for the averaged velocities i.e. Fig. 8.17.

As the radial velocity plays a crucial role in the case of rotation ratio of torque maximum it is worthy to understand the axial behaviour. In Figures 8.19, 8.20 the relative angular velocities (left) and radial velocities (right) are depicted for $Re_S = 1.65 \cdot 10^5$ and $2.19 \cdot 10^5$ respectively. The reader is directed to also see Appendix C Figures C.12 to C.6 to get the full view of all measured shear Reynolds numbers for the torque maximum. Obviously the large-scale circulation dominates the flow. At the region of the outflow the the large-scale circulation advects fluid with large angular momentum outwards and in the region of the inflow it transports low momentum inwards. Obviously this transport by the large-scale circulation is strong enough to increase the entire transport process and

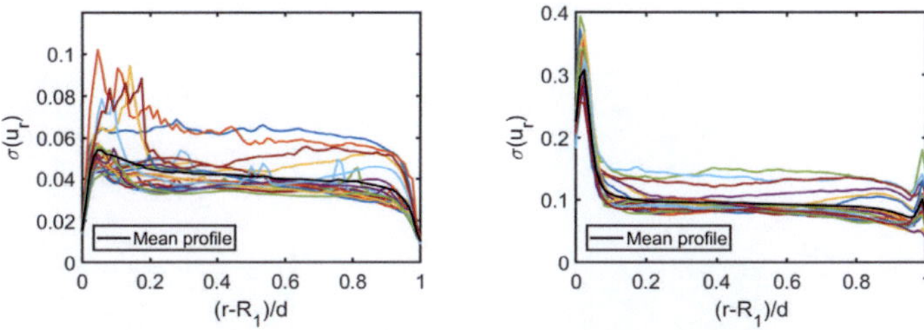

FIGURE 8.18: Profiles of the standard deviation of the radial velocity component computed in time and azimuthal coordinate for $Re_S = 80000$ (left) and $Re_S = 201000$ (right) $\mu = -0.20$. In colour the different axial positions are depicted, the black solid line represents the axial mean profile.

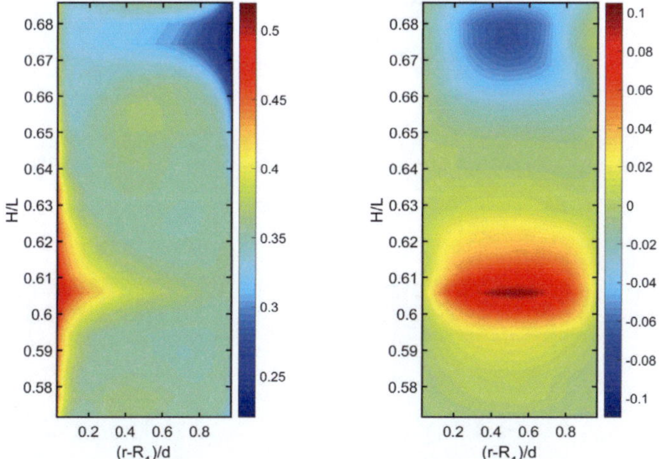

FIGURE 8.19: Contours of the Angular (left) and radial (right) velocities averaged in time and azimuthal coordinate for $Re_S = 165000$, $\mu = -0.20$.

increase the torque at the wall. In Figure 8.21 the local fluctuation of the velocities is also given for the two describes shear Reynolds numbers. It can be identified, that the outflow takes highly turbulent flow and advects it to the outer cylinder and vise versa.

It has to be mentioned, that for all cases the large-scale circulation wave length seems to be very similar. Only in the case of $Re_S = 1.65 \cdot 10^5$ the in- and outflow seem to be at different positions. As it was not possible to perform a global visualisation together with the PIV measurements, as the particles differ strongly, no information of the exact global behaviour can be done.

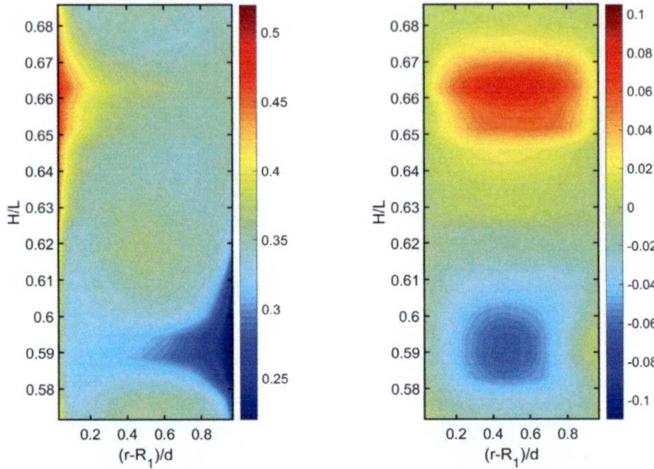

FIGURE 8.20: Contours of the Angular (left) and radial (right) velocities averaged in time and azimuthal coordinate for $Re_S = 219000$, $\mu = -0.20$.

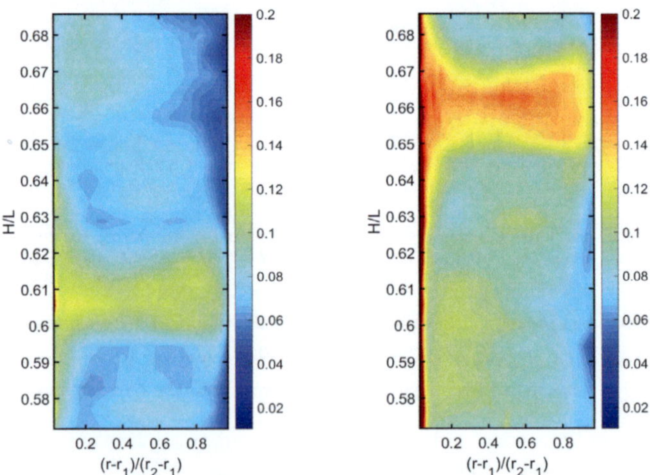

FIGURE 8.21: Contours of the fluctuation of the radial velocity for $\mu = -0.20$, $Re_S = 165000$ (left) and 219000 (right). Fluctuation calculated as the standard deviation time and azimuthal coordinate.

8.6 Influence of the rotation ratio for constant shear Reynolds number

Up to this point the Velocimetry was analysed for the cases of a fix rotation ratio μ, but varying shear Reynolds number. As there are huge differences of the flow behaviour depending on the rotation ratio in the next section this dependency is studied. For the two different shear Reynolds numbers of $Re_S = 8 \cdot 10^4$ and 10^5 the rotation ratio is varied. The flow for co-rotating cylinders $\mu = +0.20$ and $+0.10$, which is already linear-instable (Rayleigh criterion says that flow is centrifugally stable for $\mu > \eta^2 = 0.25$), is investigated as well as the pure inner cylinder rotation, which was already discussed. For the case of counter-rotation the rotation ratios have been set to $\mu = -0.10, -0.15, -0.20, -0.25, -0.30, -0.40$ and -0.50.

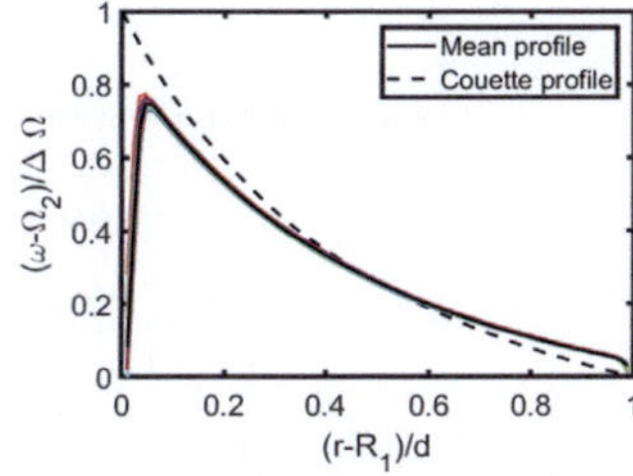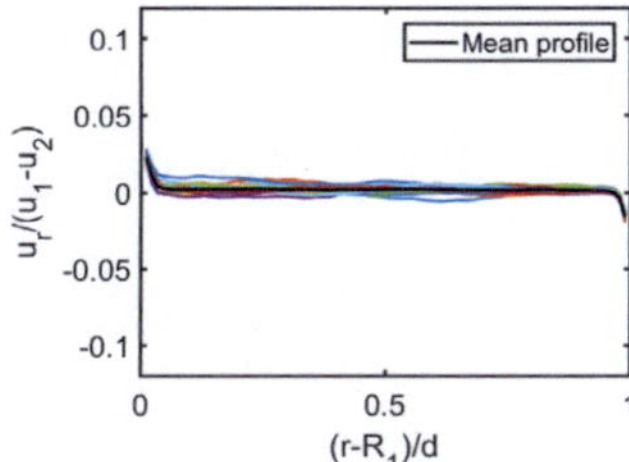

FIGURE 8.22: Angular and radial velocity profiles averaged in time and azimuthal coordinate for $Re_S = 80000$, $\mu = +0.20$. In colour the different axial positions are depicted, the black solid line represents the axial mean profile and the dashed line in angular velocity profile represents the laminar Couette-solution.

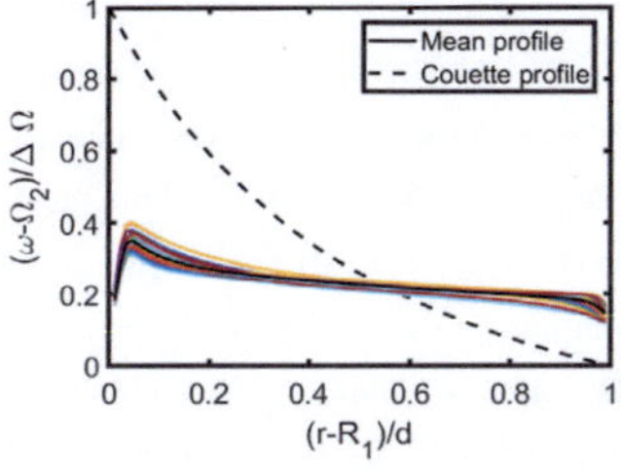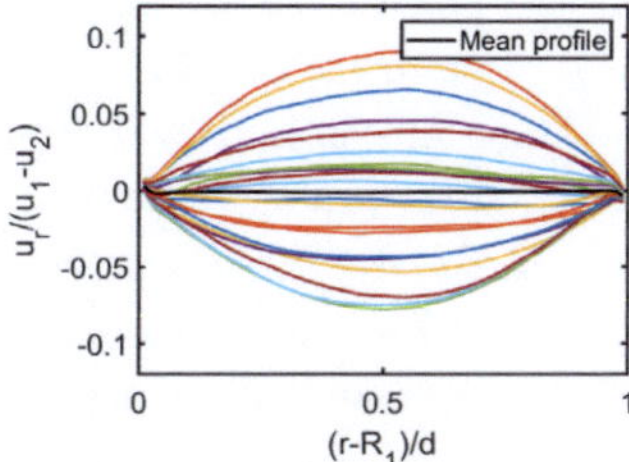

FIGURE 8.23: Angular and radial velocity profiles averaged in time and azimuthal coordinate for $Re_S = 80000$, $\mu = -0.15$. In colour the different axial positions are depicted, the black solid line represents the axial mean profile and the dashed line in angular velocity profile represents the laminar Couette-solution.

The velocity profiles for the case of $\mu = 0$ and $\mu = -0.20$ have been already discussed. For the case of co-rotating cylinders the profiles get closer to the Couette-solution. The

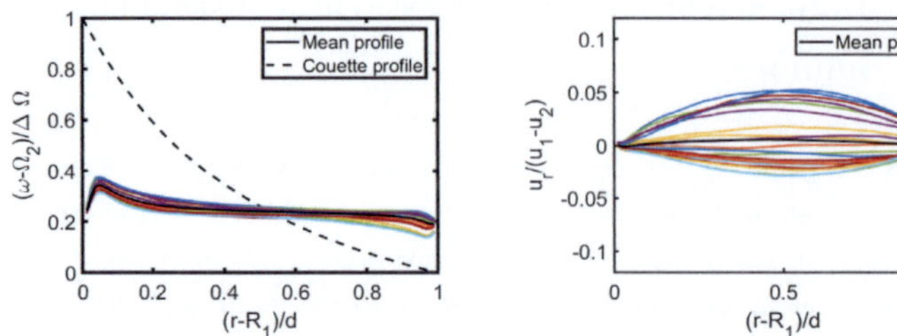

FIGURE 8.24: Angular and radial velocity profiles averaged in time and azimuthal coordinate for $Re_S = 80000$, $\mu = -0.25$. In colour the different axial positions are depicted, the black solid line represents the axial mean profile and the dashed line in angular velocity profile represents the laminar Couette-solution.

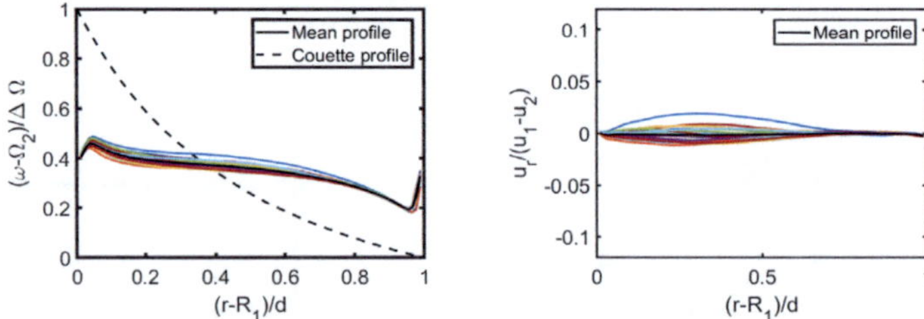

FIGURE 8.25: Angular and radial velocity profiles averaged in time and azimuthal coordinate for $Re_S = 80000$, $\mu = -0.60$. In colour the different axial positions are depicted, the black solid line represents the axial mean profile and the dashed line in angular velocity profile represents the laminar Couette-solution.

radial velocity becomes neglitible while for $\mu = +0.20$ and $Re_S = 8 \cdot 10^4$ the angular velocity profile is slightly deflected from a laminar solution (Figure 8.22). For the case of $\mu = +0.10$ the deviation from the Couette profile becomes stronger, but still the behaviour is similar to it (Figure B.27). As soon as the outer cylinder rotation is driven in the opposing direction of the inner cylinder the flow changes drastically to the one observed for $\mu = 0$ (Figure 8.9). For $\mu = -0.10$ the radial velocity sets on. The angular velocity profile gets flatter than for outer cylinder at rest. The large-scale circulation is enlarged due to the stabilizing effect of the outer cylinder rotation (Figure B.29). The strength of the large-scale circulation increases with decreasing rotation ratio ($\mu = -0.15$: Figure 8.23) until the torque maximum is reached at $\mu = -0.20$. Passing the maximum ($\mu = -0.25$) results in a decrease in the radial velocity component in Figure 8.24. Increasing the counter rotating even more the radial velocity close to the outer cylinder vanishes, while in the inner part of the system still an axial dependence of the radial velocity can be observed, as in Figure 8.25. Basically the same can be seen for the second shear Reynolds number $Re_S = 10^5$ the rotation ratio was varied. In

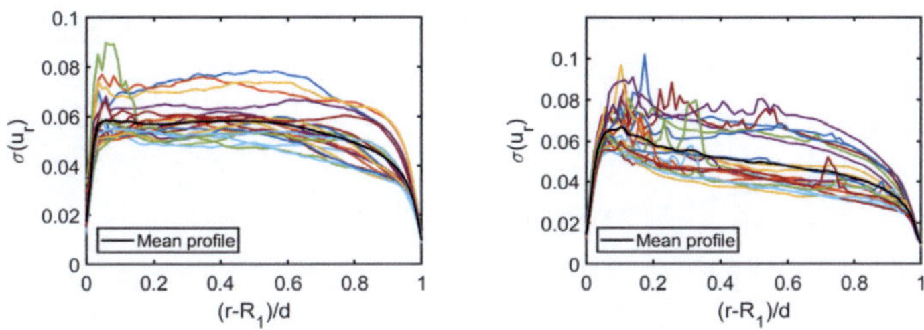

FIGURE 8.26: Profiles of the standard deviation of the radial velocity component computed in time and azimuthal coordinate for $Re_S = 80000$, $\mu = -0.15$ (left) and $\mu = -0.25$ (right). In colour the different axial positions are depicted, the black solid line represents the axial mean profile.

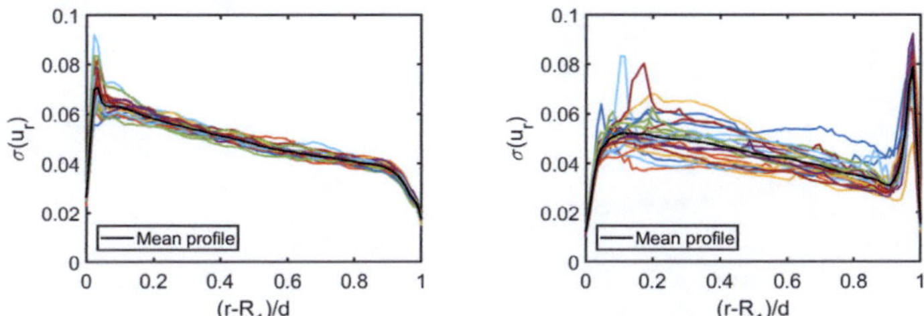

FIGURE 8.27: Profiles of the standard deviation of the radial velocity component computed in time and azimuthal coordinate for $Re_S = 80000$, $\mu = +0.20$ (left) and $\mu = -0.60$ (right). In colour the different axial positions are depicted, the black solid line represents the axial mean profile.

Figure 8.28 the low counter rotation is depicted in contrast to Figure 8.29, where the torque maximum is already passed. The entire measured velocity profiles are also shown in Appendix B Figures B.26 onto B.45.

The local velocity fluctuation of the radial velocity component also changes drastically for a constant shear Reynolds number but the different rotation ratios. In Figures 8.26, 8.27 the behaviour is given for $Re_S = 8 \cdot 10^4$ and in Figure 8.30 are two cases for $Re_S = 10^5$. For the entire profiles the reader is directed to Appendix B Figures B.58 to **??**. For the case of co-rotation a decrease of the fluctuations along the radius can be observed, while there is no axial dependency observed. The strong curved fluctuation profile as for $\mu = 0$ is changed by the flow for counter rotating cylinders. Here a flat behaviour in the bulk can be seen, while in axial direction a huge deviation of the fluctuations is given. For strong counter rotation (i.e. $\mu = -0.6$) the axial mean slope again is similar to the one of co-rotation $\mu = +0.20$, while a peak can be observed close to the outer cylinder. An axial dependency can still be observed, although it is weak.

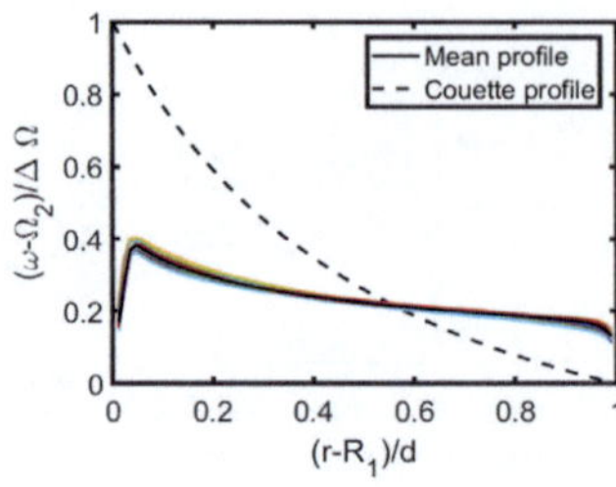
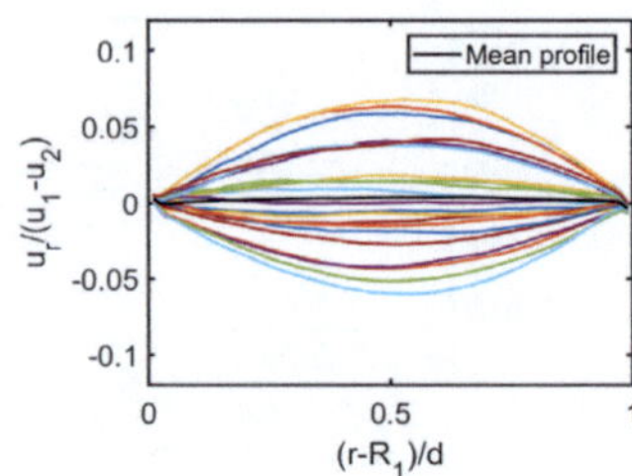

FIGURE 8.28: Angular and radial velocity profiles averaged in time and azimuthal coordinate for $Re_S = 100000$, $\mu = -0.10$. In colour the different axial positions are depicted, the black solid line represents the axial mean profile and the dashed line in angular velocity profile represents the laminar Couette-solution.

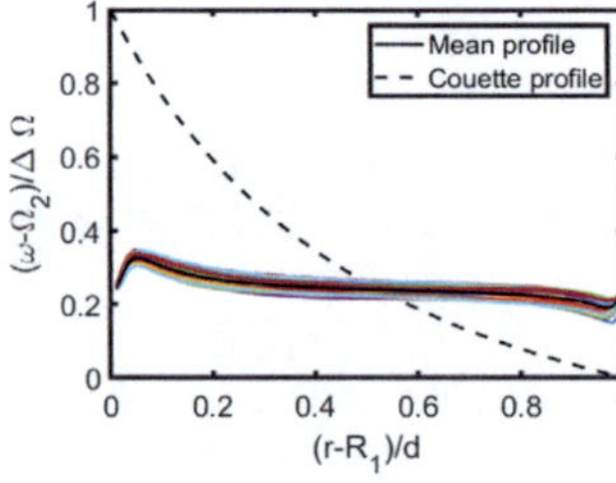
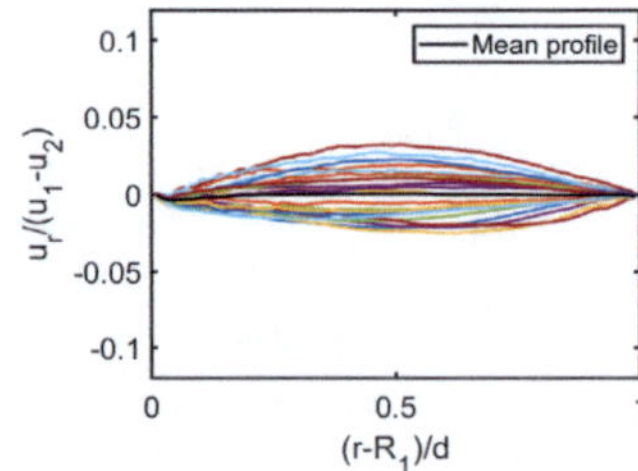

FIGURE 8.29: Angular and radial velocity profiles averaged in time and azimuthal coordinate for $Re_S = 100000$, $\mu = -0.30$. In colour the different axial positions are depicted, the black solid line represents the axial mean profile and the dashed line in angular velocity profile represents the laminar Couette-solution.

A convolute of the temporal, azimuthal and axial averaged velocity profiles for constant shear Reynolds number and varied rotation rate μ are depicted in Figure 8.37, while the deviations computed in time, azimuthal and axial direction is shown in Figure 8.38

This leads to the reconstruction of the radial velocity profiles at different heights in Figures 8.31 to 8.33 (respectively Appendix C Figures C.7 to C.26). It is quiet clear that the radial velocity component is resulting from the turbulent Taylor-vortices in the counter rotating regime. For pure inner cylinder rotation and high shear Reynolds numbers this large-scale circulation did disappear. Classically one would expect, that the counter rotation would stabilize the flow from the outers cylinder site. This does happen for the strong counter rotation, where in the outer part of the gap the system becomes less radial velocity component and the fluctuations also decrease. At the inner part of the gap an instable flow exists and is suppressed by the outer stable flow from the outer cylinder. But, the system has the ability, that the large-scale circulation can enhance into the stable layer.

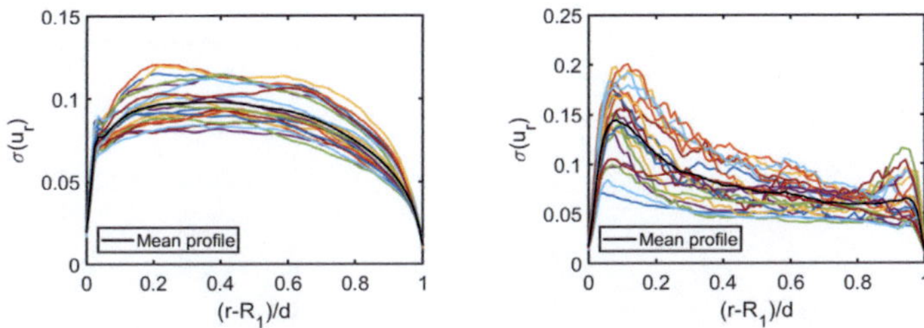

FIGURE 8.30: Profiles of the standard deviation of the radial velocity component computed in time and azimuthal coordinate for $Re_S = 100000$, $\mu = -0.10$ (left) and $\mu = -0.30$ (right). In colour the different axial positions are depicted, the black solid line represents the axial mean profile.

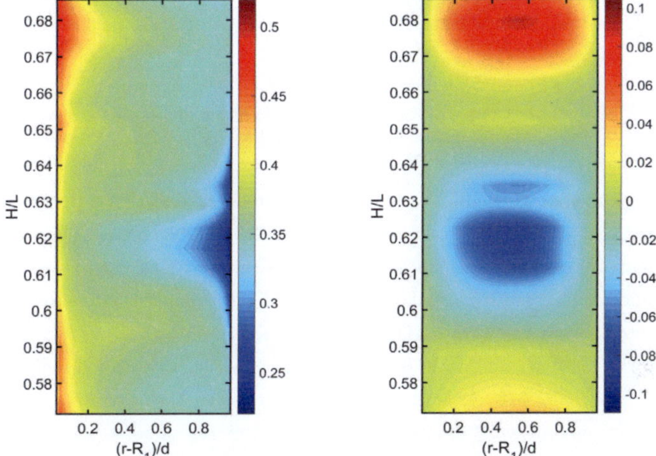

FIGURE 8.31: Contours of the Angular (left) and radial (right) velocities averaged in time and azimuthal coordinate for $Re_S = 80000$, $\mu = -0.15$.

For counter rotation the inner cylinder rotates with positive angular velocity while the sign of the outer cylinders angular velocity is opposite. Thus, there must be a radial location in the flow where the angular velocity equals zero. In the whole cylinder gap this radial location forms a surface. For the solution of inviscid fluid, this surface separates the region of linear stable flow (from neutral surface until outer cylinder) and the linear unstable region (inner cylinder to neutral surface). The position can be calculated, according to Chandrasekhar [6] by: $r_{n,inv}(\mu) = R_1(1 - \mu)^{1/2}(\eta^2 - \mu)^{-1/2}$. For the case of pure inner cylinder rotation the position is exactly at the outer cylinder wall, as this one is at rest. For the case of counter-rotation the neutral surface must be inside the

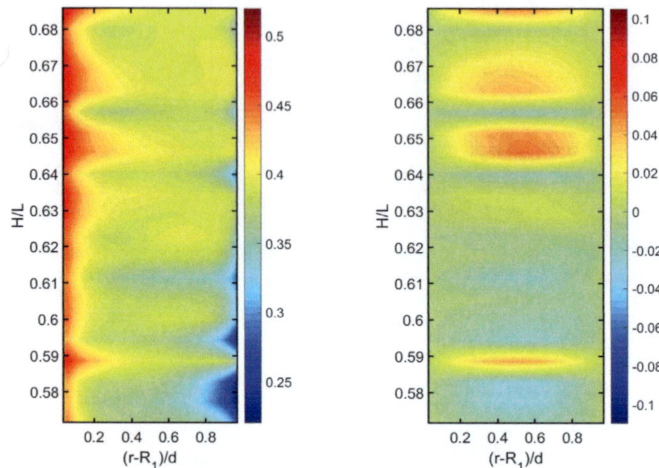

FIGURE 8.32: Contours of the Angular (left) and radial (right) velocities averaged in time and azimuthal coordinate for $Re_S = 80000$, $\mu = -0.25$.

gap. For viscous fluids, the large-scale circulation, happening in the linear unstable inner part can be enhanced into the theoretical stable region. A prediction for the extend of a vortex into a stable flow was presented by Esser and Grossmann [12], while Brauckmann and Eckhardt [3] did expressed the enhanced radius for counter rotating Taylor-Couette flow at Esser-Grossmann radius: $r_{n,EG}(\mu) = R_1 + a(\eta)(r_{n,inv} - R_1)$, where $a(\eta) = (1-\eta) \cdot ((1+\eta)^{3/2}(2+6\eta)^{(-1/2)} - \eta)^{-1}$ for the investigated radius ratio results in $a(\eta = 0.5) = 1.5548$. As soon as this extension of the Taylor-vortices hits the outer wall, a radial inhomogenity happens induced by evolving intermittency. Brauckmann and Eckhardt [3] showed especially, that this prediction can be used for the determination of the torque maximum. For pure inner cylinder rotation, the hypothetical radius $r_{n,EG}$ is outside the system. The vortices can not grow up to this point and they are suppressed. For very strong counter rotation the vortices may grow until $r_{n,EG}$, but there is still undisturbed flow surrounding it, suppressing the angular momentum transport due to its stability. Interestingly the rotation ratio, where the position of the EG-radius equals the outer cylinder radius $r_{n,EG}(\mu_{EG}) = R_2$, follows to be $\mu_{EG} = -0.191$. Obviously this intersects with the observed maximum in torque within a measurement tolerance. In Figure 8.36 the position of these radii are depicted. In addition also the positions, where the mean angular velocity equals zero, calculated from the PIV measurements are drawn in. As in axial direction the neutral surface deviates, the shaded region indicates the largest amplitudes, the neutral location was computed.

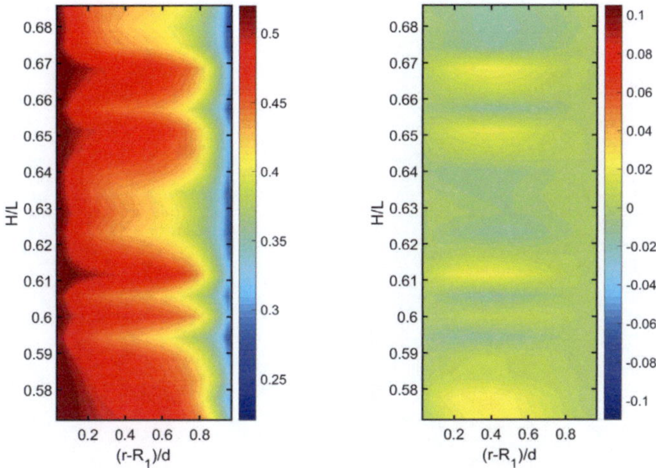

FIGURE 8.33: Contours of the Angular (left) and radial (right) velocities averaged in time and azimuthal coordinate for $Re_S = 80000$, $\mu = -0.40$.

Both shear Reynolds number intersect well in this graph. For the flow in the regime between pure inner cylinder rotation and torque maximum, the computed locations always lay at the outer cylinder wall, within the resolution of this measurement. As soon as the rotation ratio passes the torque maximum the detachment of the neutral surface can be observed. This is in well agreement with the prediction of Brauckmann and Eckhardt [3]. The detachment starts at the inflow regions of the large-scale circulation as shown in Figure 8.36 right. For higher counter rotation the variations of the measurements become larger and the mean value also leave the prediction given by Brauckmann and Eckhardt [3] and reach the inviscid solution for the rotation ratio $\mu = -0.60$

The theory of the enhanced large-scale circulation was derived by Brauckmann and Eckhardt [3] explaining the observed torque maximum and flow behaviour. They had only the possibility to study the flow for relatively low shear Reynolds numbers up to $Re_S = 2 \cdot 10^4$. But there results where supported by the torque measurements of this study, published in Merbold et al. [24] for large shear Reynolds numbers. With the present measurements of the fluid flow in the gap this theory can be confirmed. The large-scale circulation does grow into the hypothetical stable flow. Thus, the entire gap becomes instable for weak counter rotation $0 >= \mu >= -0.20$. This enhancement of the large-scale circulation transports very efficient the angular motion and even suppresses turbulent fluctuations existent for the pure inner cylinder rotation.

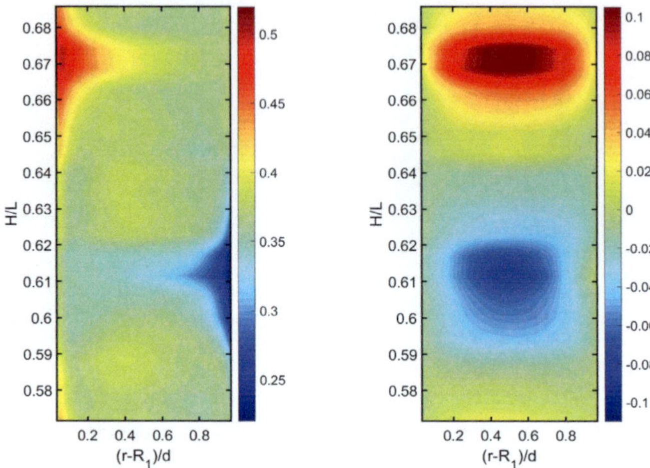

FIGURE 8.34: Contours of the Angular (left) and radial (right) velocities averaged in time and azimuthal coordinate for $Re_S = 100000$, $\mu = -0.20$.

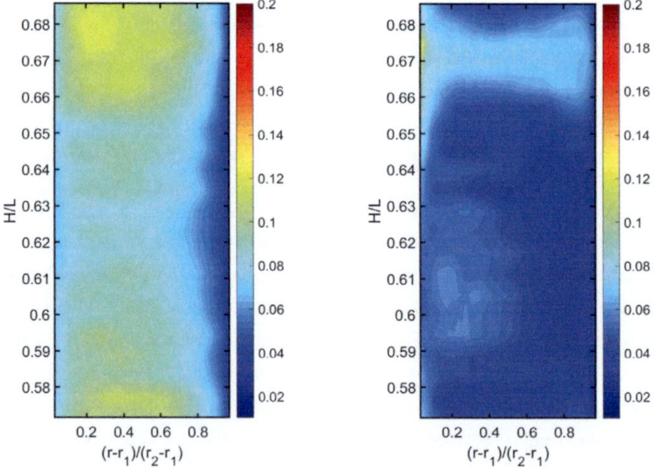

FIGURE 8.35: Contours of the fluctuation of the radial velocity for $\mu = -0.10$ (left) and $\mu = -0.20$ (right) at $Re_S = 100000$. Fluctuation calculated as the standard deviation time and azimuthal coordinate.

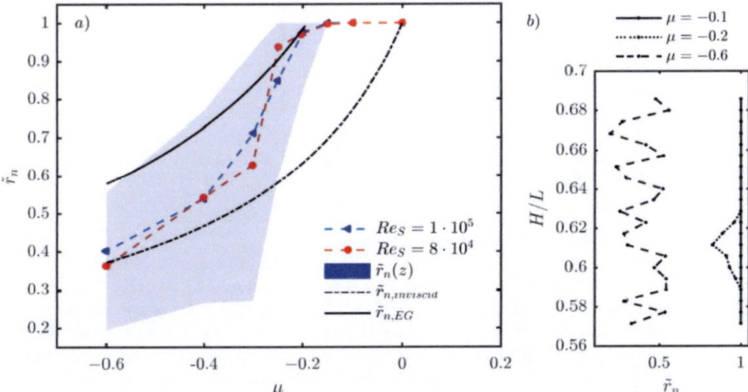

FIGURE 8.36: a) Radial location of the averaged neutral surface r_n for $Re_S = 80000$ and 100000 depending on the rotation ratio μ. The positions are averaged in time, azimuth and axial direction, while the shaded area indicates the amplitudes of the neutral surface along the height of the system. (b) The neutral surface as a function of height for $Re_S = 100000$.

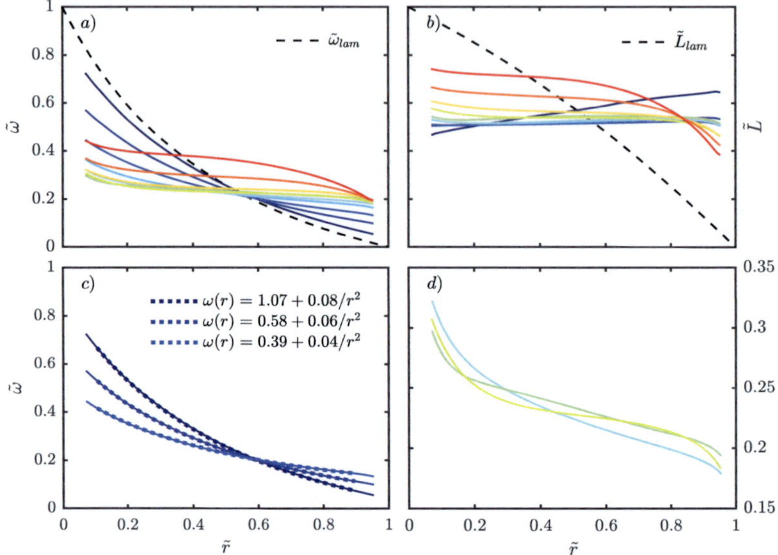

FIGURE 8.37: Angular velocity (a,c,d) and angular momentum (b) profiles of the temporal, azimuthal and axial averaged fields for constant shear Reynolds number $Re_S = 100000$. The angular velocity is normalized as $\tilde{\omega} = (\omega - \Omega_2)/\Delta\Omega$ and the angular momentum respectively $\tilde{L} = (L - L_2)/(L_1 - L_2)$. The colours are the same as in Figure 8.38. The dashed line represents the laminar Couette solution. The dotted lines in (c) represent bulk profiles based on laminar Couette solution, where the boundaries at $\omega_{1,i}(\tilde{r} = 0.1)$ and $\omega_{2,i}(\tilde{r} = 0.9)$ as derived for velocity profiles in Chapter 5.

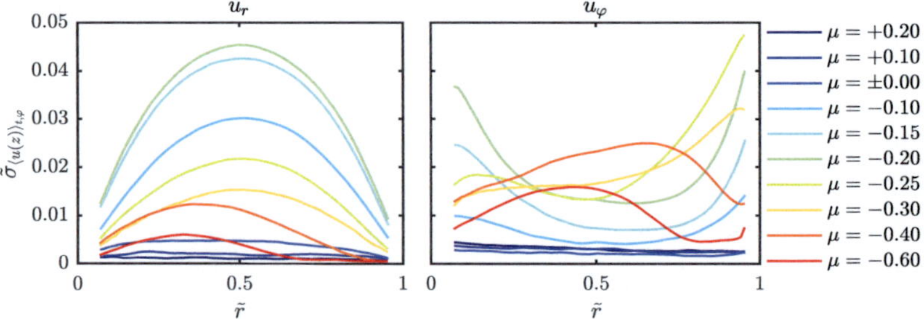

FIGURE 8.38: Axial standard deviation profiles of the radial (left) and azimuthal velocity (right) averaged in time and azimuthal direction for constant shear Reynolds number $Re_S = 100000$ and varying rotation ratio μ.

8.7 Energy distribution

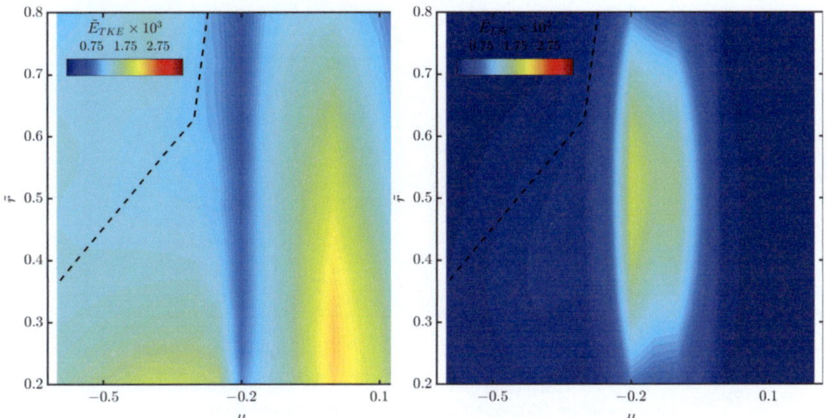

FIGURE 8.39: Contour plots of the temporal and spatial averaged (t, φ, z), normalized kinetic energy fractions at $Re_S = 80000$ of the turbulent kinetic energy E_{TKE} and the kinetic energy of the large-scale circulation E_{LSC} as a function of the gap position and the rotation ratio μ. The dashed line represents the location of the neutral surface.

From the flow measurements it is possible to compute the kinetic energy of the flow. This offers an insight into the changes of the flow state more prominent than the flow velocities itself. Using $u = \langle u \rangle_{\varphi,t} + u' = \bar{u} + u'$, representing a classical Reynolds averaging, results for the kinetic energy averaged over time and space (t, φ, z):

$$E_{kin}(r) = \frac{1}{2} \langle \bar{u}_\varphi^2 + \bar{u}_r^2 + \bar{u}_z^2 + u_\varphi'^2 + u_r'^2 + u_z'^2 \rangle_{t,\varphi,z} \tag{8.1}$$

The mixed terms do vanish and the first term represents the energy of the driving base flow in azimuthal direction and not part of the secondary flow, transporting momentum. The flow measurements did not allow to measure the axial velocity component by the single camera in each light sheet. Thus, we have to neglect the u_z component. It remains the radial velocity component of the mean secondary flow as well as the turbulent fluctuations. This leads to two terms of kinetic energy, the one of the large-scale circulation E_{LSC} and the turbulent kinetic energy in azimuthal and radial direction E_{TKE}:

$$E_{LSC}(r) = \frac{1}{2} \langle \bar{u}_r^2 \rangle_{t,\varphi,z} \qquad E_{TKE} = \frac{1}{2} \langle u_\varphi'^2 + u_r'^2 \rangle_{t,\varphi,z} \tag{8.2}$$

The sum of the both is defined as the total kinetic energy $E_{tot}(r) = E_{LSC}(r) + E_{TKE}(r)$ and being normalized using the shear velocity u_s to $\tilde{E} = E/(u_S^2/2)$. The result of these calculations are depicted in Figure 8.39 left for the turbulent kinetic energy, in Figure

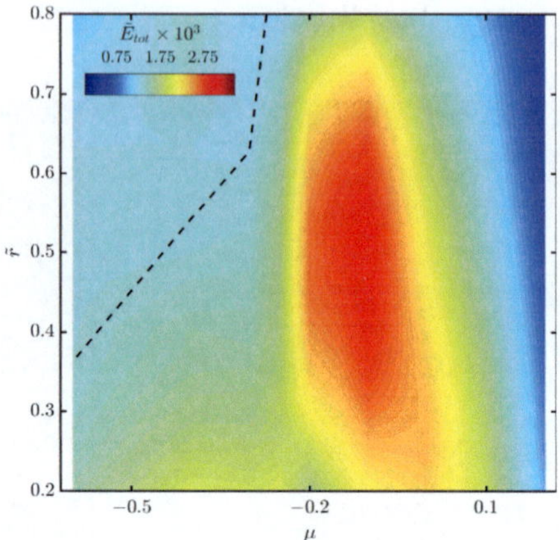

FIGURE 8.40: Contour plot of the temporal and spatial averaged (t, φ, z), normalized kinetic energy fractions at $Re_S = 80000$. E_{tot} represents the sum of the turbulent kinetic energy E_{TKE} and the kinetic energy of the large-scale circulation E_{LSC} as a function of the gap position and the rotation ratio μ. The dashed line represents the location of the neutral surface.

8.39 right for the large-scale circulation and in Figure 8.40 for the total of this both contributions in dependence of the relative gap position $\tilde{r} = (r - R_1)/d$ and the rotation ratio μ. In addition the dashed line indicates the calculated position of the angular velocity averaged in time, azimuthal and axial direction becomes $\langle \omega(r) \rangle_{t,\varphi,z} = 0$, the so called neutral velocity.

For the inner gap region the TKE is clearly larger than for the outer gap region (Figure 8.39 left). Looking to the behaviour along the rotation ratio it becomes clear, that the TKE is very strong for outer cylinder at rest ($\mu = 0$). For co-rotating cylinders the TKE decreases to very low values, close to a laminar flow. In the regime between pure inner cylinder rotation and the torque maximum the TKE does decrease. Finally at the rotation ratio of the torque maximum, the TKE shows up a very low value. This illustrates that the formation of the large-scale circulation goes in parallel with a more coherent flow state. As the rotation ratio decreases to the high counter-rotating regime now the neutral velocity detaches from the outer wall and tends into the gap. At this area the TKE is also at low values, while in the outer and inner layer the TKE is slightly larger for those rotation ratios. Close to the inner cylinder a huge impact of TKE can be observed in a narrow region. This increase can be caused by the intermittency described by Brauckmann and Eckhardt [3].

The energy of the large-scale circulation E_{LSC}, Figure 8.39 right, for co-rotation is basically at zero. Also for the strong counter rotation ($\mu \leq -0.4$) there are only very

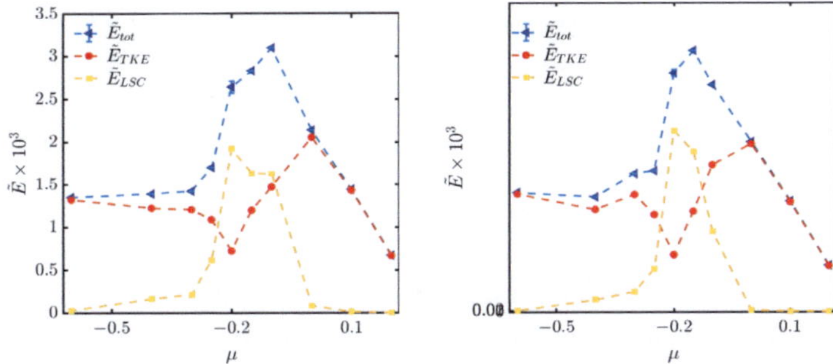

FIGURE 8.41: Fractions of the kinetic energy of the turbulence (E_{TKE}), large-scale circulation (E_{LSC}) and the sum of both (E_{tot}) averaged in time, azimuthal and axial direction in dependence of the rotation ratio μ at radial mid gap for $Re_S = 80000$ (left) and 100000 (right).

small values computed. Just close to the inner cylinder a small deviation from zero can be seen. Obviously the E_{LSC} forms a clear maximum at the rotation rate of the torque maximum over the whole gap. This maximum is also visible in the other low counter rotating cases ($\mu = -0.10$ and -0.15) up to the pure inner cylinder rotation, where the E_{LSC} vanishes. The decrease to stronger counter rotation than the torque maximum happens suddenly.

The total energy E_{tot} gives the final image of the turbulent and large-scale transport, depicted in Figure 8.40. The maximum of the large-scale circulation at the torque maximum dominates and forms also the maximum in the total kinetic energy. For stronger counter rotation E_{tot} decreases as the kinetic energy of the large-scale circulation drops more than the TKE increases. For the cases of lower counter rotation ($\mu = -0.15$) the total energy seem to be very similar as for torque maximum, while the small decrease of the E_{LSC} is compensated by an increase of the E_{TKE}. Going further the decrease of $E_{L}SC$ also affects the total kinetic energy to decrease.

The axial velocity was not measured during this campaign. Thus, in the energy budget of the kinetic energies it is missing. Assuming that the axial velocity component of an idealized turbulent Taylor-vortex vanishes for the centre of the gap ($\tilde{r} = 0.5$) the axial component should not matter in the energy of the large-scale circulation at this position. Thus in Figure 8.41 the energies at the radial position of the mid gap are depicted for the shear Reynolds numbers $Re_S = 8 \cdot 10^4$ and 10^5. The whole flow behaviour can be summarized y that.

First of all, at both shear Reynolds numbers investigated the behaviour of the kinetic energies is similar. In both cases the energy of the large-scale circulation is neglictible for co-rotating system as well as for outer cylinder at rest, $\mu \geq 0$. Here the total kinetic energy is driven by the turbulence. As the counter rotation sets on E_{LSC} becomes more dominant than E_{TKE}. The TKE already decreases with the counter rotation in contrast to outer cylinder at rest. For the large $Re_S = 10^5$ still the turbulent kinetic energy is larger than the one of the large-scale circulation. For $Re_S = 8 \cdot 10^4$ they are contributing nearly the same to the transport. Clearly seen can be that until the rotation ratio of the torque maximum the E_{LSC} increases and is maximized at the torque maximum, while E_{TKE} decreases and results in a minimum at the torque maximum. Passing the torque maximum drops the energy of the large-scale circulation and increases slightly the one of the turbulence. Thus the transported energies show there maximum close to the maximum.

It has to be mentioned that the wave lengths of the vortices were not constant over all measurements. Especially for shear Reynolds number $Re_S = 8 \cdot 10^4$ the different sizes were already discussed. For example a comparison of Fig. B.29 and B.31 shows these differences. Especially the case of $Re_S = 8 \cdot 10^4$ and $\mu = -0.10$ has a very different wavelength compared with the other measurements. It is known that different numbers of vortices affect the resulting transport Martinez-Arias et al. [22]. Thus, the deviation of this case from the others in terms of the energy contributions can be explained.

It is to be noted, that in this analysis the axial velocity did not played a role and time dependent instabilities, such as the wavy patterns, observed in the flow visualisation in Chapter 6 are excluded from the analysis. To catch the these, a time resolved measurement including all three velocity components would be required. It could be, that using a secondary flow measurement technique in parallel could reveal the axial time dependent flow and allow for a phase-averaging. But such measurements were not possible so far.

8.8   Nusselt number from flow fields

In Chapter 4 the angular momentum transport in the Taylor-Couette system, represented by the quasi Nusselt number Nu_ω, is measured by means of direct torque measurements. In the following the flow behaviour was analysed by different measurements. Flow visualisations have revealed the global large-scale circulation for counter-rotating flow, which was analysed by the PIV in azimuthal-axial planes and also by the use of the azimuthal planes traversed along the height of the system. By the last of these measurements the Nusselt number can be directly computed from the flow fields. Moreover, by the flow fields the contributions of the flow field to the Nusselt number can be quantified (see also Brauckmann et al. [4], Froitzheim et al. [13]). Using the prior used flow field decomposition $(u = \bar{u} + u')$, leads to:

$$Nu_\omega = Nu_\omega^{turb} + Nu_\omega^{LSC}, \tag{8.3}$$

with the contributions of the turbulence (turb) and large-scale circulation (LSC) averaged in radial direction:

$$Nu_\omega^{turb} = J_{lam}^{-1} \langle r^3 \langle u_r' \omega' \rangle_{A(r),t} \rangle_r \tag{8.4}$$

$$Nu_\omega^{LSC} = J_{lam}^{-1} \langle r^3 (\langle \bar{u}_r \bar{\omega} \rangle_{A(r),t} - \nu \partial_r \langle \bar{\omega} \rangle_{A(r),t}) \rangle_r \tag{8.5}$$

The mixed terms $\langle \bar{u}_r \omega_r' \rangle_{A(r),t}$, $\langle u_r' \bar{\omega} \rangle_{A(r),t}$ and $\partial_r \langle \omega' \rangle_{A(r),t}$ are linear in the deviation quantities and do vanish. Thus, in the turbulent contribution to the Nusselt number it remains the Reynolds stress term and the large-scale circulation contribution to the Nusselt number consists of the gradient of the mean angular velocity and the product of the averaged radial and angular velocities. When the spatial average across cylindrical surfaces $A(r)$ is performed along the whole length, the radial velocity component averages out. Thus, this term can be neglected. The radial profiles of the computed Nusselt numbers for the measured Re_S and outer cylinder at rest ($\mu = 0$) are depicted in Figure 8.42. By definition the Nusselt number has to be constant in respect to the radial position, the computed Nusselt numbers from these measurements are relatively constant in radial direction, which is in good agreement. Close to the cylinder walls the gradient of the angular velocity becomes large and the spatial resolution does not allow an accurate estimation, so these regions have to be taken out of interest and an interval of $\tilde{r} \in [0.1, 0.9]$ remains. To finally compute the Nusselt number for the set flow, it is averaged along the radius. The radial behavior in addition is shown in Figure 8.42 right compensated by the radially averaged Nusselt numbers. A trend along the radius can be observed with a positive slope along the radius. Close to the inner cylinder the calculated Nusselt numbers are smaller than close to the outer cylinder. This indicates, that that the Nusselt number is radially underestimated. The reasons for this lay in the resolution in time and space to estimate the Reynolds stress term $\langle u_r' \omega' \rangle_{A(r),t}$, which has the leading contribution to the total Nusselt number. Close to the inner cylinder the velocities are the largest. Thus, the displacements between particle images are larger, while the fluctuations are also large. Hence, there computation is limited by the temporal and spatial resolution (depending on axial plane: $0.379 - 0.46mm$). Thus, the underestimation is more dominant close to the inner cylinder, explaining the slope in radial direction.

In Figure 8.43 the Nusselt numbers from the PIV are compared to the Nusselt numbers directly measured with the torque sensors and discussed in Chapter 4 and direct numerical simulations done by Ostilla-Mónico et al. [27, 28]. It can be seen, that overall a slight underestimation of the Nusselt number takes place for all shear Reynolds numbers. The deviation in comparison to the measured torques lay between 19% for $Re_S = 5 \cdot 10^4$ up to 36% for $Re_S = 2.2 \cdot 10^5$. In contrast it also shows the quality of the performed measurements, as the Nusselt number is nearly completely resolved, apart from a turbulent proportion of small scales and the DNS of Ostilla-Monico resolves the full scales and is in well agreement with the torque measurements.

The scaling of the torque already, discussed in Chapter 4, can be also adopted to these data. Despite the huge number of PIV measurements along Re_S, it is still a small number to compute a scaling law based in it. In Figure 8.43 (b) the Nusselt number data is compensated by the power-law Ansatz of $Nu_\omega \sim Re_S^\beta$ using the exponent β computed for similar shear Reynolds numbers of $Re_S = 10^5$ (compare Figure 4.2). Figure 8.43 (b) reveals the change in the scaling clearly for the direct torque data, but only blurred for the PIV results and the numerical simulations. Fitting the PIV data to the same Ansatz the resulting exponent is $\beta_{PIV,CR} = 0.63 \pm 0.11$ for the classical regime and $\beta_{PIV,UR} = 0.76 \pm 0.08$ for ultimate regime. The results of torque data ($\beta_{Tor,CR} = 1.62 \pm 0.04$, $\beta_{Tor,UR} = 1.78 \pm 0.06$) and the Particle Image Velocimetry finally intersect within there confidential intervals. Additionally Ostilla-Mónico et al. [27, 28] report $\beta_{DNS,CR} \approx 0.66$, $\beta_{DNS,UR} \approx 0.76$ independent of the radius ratio. Experimental data of two different measurement techniques as well as the numerical approach finally show good agreement with each other.

Apart from the dependence of the Nusselt number with shear Reynolds number for outer cylinder at rest, also the measurements for varying rotation ratio and fixed shear Reynolds number have to be taken into account. The decompensation derived in Equation 8.3 is performed for $Re_S = 10^5$ and the different situations of rotation ratio. In Figure 8.44 (a) the radial profiles of Nu_ω lead to the observation, that for these cases the radial deviations become even larger in contrast to $\mu = 0$. The radially averaged Nusselt number Nu_ω is depicted in Figure 8.44 (b) and is also compared to the direct torque measurements for the same parameters. Additionally the terms computed from the PIV for the large-scale circulation Nu_ω^{LSC} as well as the turbulent mixing Nu_ω^{turb} are depicted in the same diagram.

The closer examination of the velocity profiles and the energy contributions can be finally concluded with the contributions to the Nusselt number. The angular momentum flux due to the large-scale circulation for pure inner cylinder rotation ($\mu = 0$) and large shear Reynolds numbers is approximately zero. Nu_ω^{LSC} increases strongly for low counter-rotating regime and reaches finally its maximum at $\mu = -0.20$. For stronger counter-rotation the decrease takes place, but is not as steep as seen for $E_L SC$. The behaviour of the turbulent contribution to the Nusselt number Nu_ω^{turb} increases from co-rotation

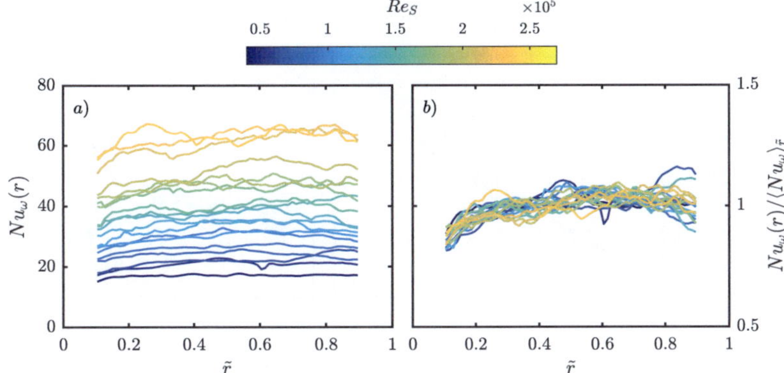

FIGURE 8.42: (a) Profiles of the quasi Nusselt number calculated from the PIV for outer cylinder at rest ($\mu = 0$) and different shear Reynolds numbers. (b) The radial behaviour of the Nusselt numbers compensated by the radial mean Nusselt number in the interval $[0.1\,0.9]$ across the gap.

towards $\mu = 0$ and even into weak counter rotation $\mu = -0.10$. A pronounced minimum of Nu_ω^{turb} is located for the case of $\mu = -0.20$, where the torque is maximized. For stronger counter rotation the contribution of $Nu_\omega^{turb}(\mu \leq -0.3)$ is relatively constant while the $Nu_\omega^{LSC}(\mu \leq -0.3)$ decreased.

In total the Nusselt number Nu_ω also maximizes at $\mu = -0.20$. The comparison with the direct torque measurements shows nearly the same results, for $\mu < 0$ the Nu_ω from PIV are slightly larger than the torque data. As already explained, the turbulent fraction must be underestimated due to the temporal and spatial resolution of these measurements, while the contribution of the large-scale circulation seems to be overestimated. This especially is visible at the torque maximum, where even a peak results in the total calculation of the Nusselt number. Both results are taken within different experimental facilities, different end plate dynamics and totally different measurement techniques (direct torque in TTCC with end plates at rest and PIV in TvTCC with endplates attached to outer cylinder) leading to the conclusion, that the agreement of the results is excellent.

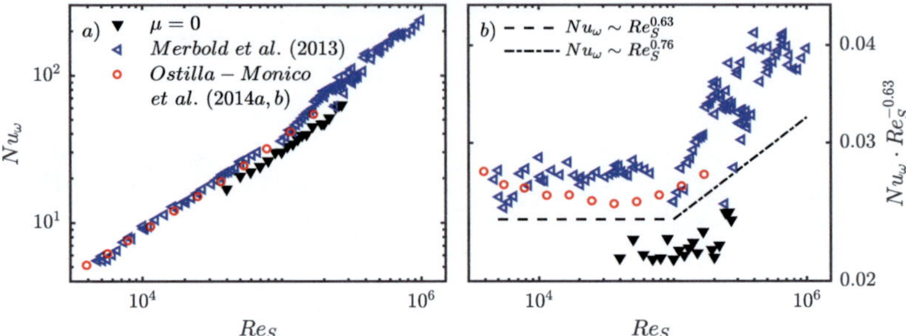

FIGURE 8.43: (a) Quasi Nusselt numbers Nu_ω in dependence of shear Reynolds number for pure inner cylinder rotation ($\mu = 0$). The black triangles represent the computed Nusselt numbers from the PIV, averaged on interval $[0.10.9]$ across the gap. Blue triangles represent the torque measurements from Chapter 4, published also in Merbold et al. [24]. Red circles represent numerical simulations by Ostilla-Mónico et al. [27, 28] for comparison. (b) The Nusselt number compensated by Reynolds number power law with exponent $\beta = 0.63$, which holds for the classical regime as a function of the shear Reynolds number. The dashed and dashed-dotted line represent least-squares fits for the classical, where β computes to 0.63, and the ultimate regime with $\beta = 0.76$.

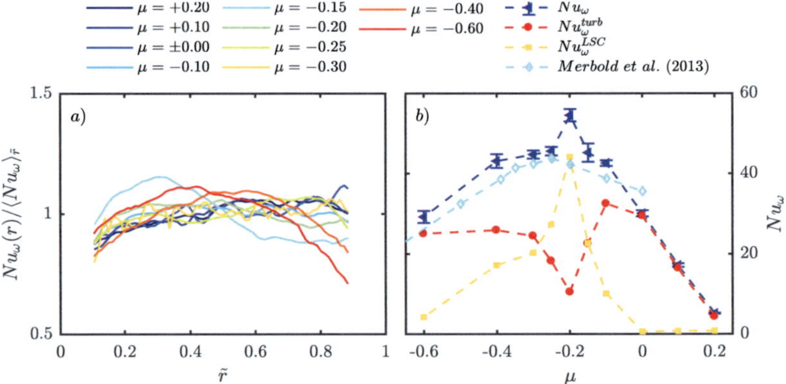

FIGURE 8.44: (a) Radial profile of the Nusselt number computed from PIV for constant shear Reynolds number $Re_s = 100000$ and different rotation ratios. (b) Nusselt number Nu_ω computed from PIV, decomposed into the contribution of the large-scale circulation Nu_ω^{LSC} and turbulence Nu_ω^{turb} as a function of the rotation ratio μ for $Re_S = 100000$. Cyan diamonds represent the torque measurements from Chapter 4, published also in Merbold et al. [24]

Chapter 9

Conclusion

9.1 General Conclusion

The aim of this thesis was to experimentally investigate the turbulent Taylor-Couette flow for radius ratio $\eta = 0.5$ at high shear Reynolds numbers and rotation of inner as well as outer cylinder. To achieve this a number of milestones were achieved.

First of all, two experimental facilities were provided to investigate the turbulent Taylor-Couette flow of wide gaps and high Reynolds numbers. While the basic system of the Turbulent Taylor-Couette Cottbus (TTCC) experiment was already existing, it needed a fundamental re-engineering. The systems rotation of the outer cylinder as well as the end plates were added to the system and the inner cylinder rotation overworked to reach the desired rotation rates of $n_1 = 5600 rpm$ to study shear Reynolds numbers above one million. It was also requested to directly measure the angular momentum flux by means of torque sensing. Thus, the inner cylinder had to be exchanged with an installed instrumentation to measure the torque forcing the inner cylinder wall. The new cylinder was split into three segments, an internal balance using strain gauges constructed, installed and calibrated and the wiring using a high-performance slip ring system realized. For all these changes of course the sealing had to be ensured. As the systems rotation number of inner cylinder increased by a factor of 15 and the outer cylinder rotation was added, the systems heat production also plaid a crucial role. To keep the isothermal condition of the system, a strong cooling system had to be taken into account. All these tasks were successfully accomplished. Due to the limitations of the optical access to the TTCC, a second experiment was designed and setup, the Top-view Taylor-Couette Cottbus experiment (TvTCC). The systems geometrical dimensions were taken over from the TTCC, but its main feature is the transparent top plate to enable sophisticated flow velocity measurements. The systems bearing had to be specially organized to keep open space on top of the experiment. The systems rotation rates do not reach the ones of the TTCC setup, but it still reaches shear Reynolds numbers of $Re_S = 4.3 \cdot 10^5$ for low viscid silicone oil. The bearing and sealing were designed to minimize friction and reduce the heat introduced into the system. Thus, no active cooling is needed to keep isothermal conditions. Also the system was made to be flexible to use different inner cylinders to allow varying the radius ratio for a wide range $(0.1 \leq \eta \leq 0.8)$. All these technical challenges had to be achieved for the following investigations of the turbulent Taylor-Couette flow.

For the present investigation fundamental scientific questions were raised on the topic of turbulent Taylor-Couette flow.

- *How large is the angular momentum flux for given flow conditions?*

- *Does the angular momentum flux scale with Reynolds number comparable with other canonical flows?*

- *Do Taylor-vortices exist for large Reynolds numbers?*

- *How does the counter-rotation affects the turbulent Taylor-Couette flow?*

The scope of the experimental study was also limited. The main limitation is the given radius ratio $\eta = 0.5$, which takes place for the whole investigation. In parallel also investigations for other radius ratios were performed for more narrow gap: $\eta = 0.71$ investigations were made but have been excluded from this thesis. For wider gaps $\eta < 0.5$ the flow is also analysed and will be the scope of future studies. The shear Reynolds numbers were limited by the maximal driving velocities and reached up to $Re_S < 1.3 \cdot 10^6$. Also the investigation had the scope for centrifugally instable flows, so the behaviour in the Rayleigh-stable regime is not studied.

A main success of the study is the precise measurement of the angular momentum flux by the use of the torque sensors installed into the inner cylinder (Chapter 4). The measurements could be realized for a wide range of shear Reynolds numbers ($10^3 < Re_S \leq 1.3 \cdot 10^6$) using silicone oils with different viscosities. Thus, the angular momentum flux could be quantified for pure inner cylinder rotation, co-rotation as well as the counter-rotating regime, the behaviour is discussed in detail in Chapter 4. Summarizing it can be expressed, that for pure inner cylinder rotation no clear power-law scaling is observed. Up to shear Reynolds number of $Re_S \approx 8 \cdot 10^4$ the exponent lies at about $\beta = 1.62$ while for larger shear Reynolds numbers β increases slightly with the shear Reynolds number while a convergence to a value of 1.78 can be suspected. It can be concluded , also supported by further experiments of the flow, that until $Re_S = 8 \cdot 10^4$

the Taylor-Couette flow of radius ratio $\eta = 0.5$ consists of a turbulent bulk flow, while the boundary layers are not turbulent - the so called classical regime (CR). Above this critical shear Reynolds number the boundary layers become fully turbulent and the system reaches the so called ultimate regime (UR). Interestingly the scaling exponent does agree well with the scaling exponent of the heat Nusselt number in Rayleigh-Bénard convection for the ultimate regime. Hence, an analogy of both systems is conducted.

As the outer cylinder rotation comes in addition, the torque signal reveals an astonishing behaviour. When the outer cylinder rotates in the opposing direction than the inner cylinder, the torque for constant shear Reynolds number increases. As usually the outer cylinder rotation stabilizes the flow it would be expected, that with adding the counter rotation the flow is also stabilized and the angular momentum flux reduced. For very strong counter rotation this expectation will be fulfilled by the measurements. The result is a maximum in angular momentum flux for rotation ratio of $\mu = -0.20$ for large shear Reynolds numbers. A comparison with direct numerical simulations by Brauckmann and Eckhardt [3] shows very good agreement for $Re_S = 5 \cdot 10^3$ up to $2 \cdot 10^4$. Here the resulting torque maximum is at lesser counter rotation. The DNS showed that for lower shear Reynolds numbers the maximum lays at the expected pure inner cylinder rotation and moves with the shear Reynolds number towards counter rotation. The torque signals for the very high Reynolds numbers finally conclude, that the maximum remains constant after $Re_S = 2 \cdot 10^4$. Along the increasing driving the maximum becomes more dominant. Nevertheless in a compensated plot it becomes obvious, that for $Re_S > 1.5 \cdot 10^5$ the peak behaviour does not change any more. For the co-rotating regime the torque values first decrease but finally rise up in the centrifugally stable regime. This is also contradictionary to what one would expect. As the centrifugally stable regime is not in the scope of this thesis, it was not analysed in more detail.

Thus the investigation itself leads to the formulation of new questions:

- *What drives the torque to be maximal for counter rotation?*

- *Which flow structures dominate the counter rotation?*

- *Can the torque be predicted by a formula?*

Further measurements were planned to answer these question. The last of them had an interest fro pure industrial applications. If there is a machinery with rotating cylinders, the torque is important to know the resulting power losses. In Chapter 4 the torque data is analysed and finally a prediction based on empirical findings is revealed with a tolerance of maximal 10% compared to the measured torques.

With the help of LDA and PIV the velocity profiles across the gap have been measured (Chapter 5). For increasing shear Reynolds number and $\mu = 0$ the normalized profiles behaved in the same way. For varying the rotation ratio the profiles changed strongly. By the first measurements only mean profiles at a few positions were possible and a more sophisticated measurement was needed. A collaborative PIV campaign inside the European High Performance Infrastructures in Turbulence (EuHIT) framework, using the high-speed PIV system of the group in Twente and the Top-view Taylor-Couette experiment could lead to a better understanding of the flow for pure inner cylinder rotation, but moreover for the torque maximum. Here strong vortices were observed for the torque maximum. The results are published in van der Veen et al. [36] and summarized also in Chapter 5.

To understand the fluid flow in the gap between the rotating cylinders it was ideally suited to perform flow visualisations. By the use of Kaliroscope particles the vortices could easily be detected (Chapter 6). The technique works well even for highly turbulent situations. It has to be mentioned, that by eye an experimentator can grab the situation of the flow far better, than what a camera can capture finally. As the turbulent Taylor-Couette flow is time dependent the captured pictures had to resolve the temporal behaviour, so videos were captured. In the flow visualisation a dye was used to only see the flow behaviour close to the outer wall and give the camera a strong contrast to capture the flow.

Using this setup also a Particle Image Velocimetry was performed for the azimuthal-axial curved plane illuminated by a strong light source outside the system (Chapter 7). Due to the dye the view was limited to only $1 - 2mm$ into the gap. This effective sheet thickness is comparable with common PIV measurements using planar light sheets from a LASER. Thus, the azimuthal and axial velocity components were analysed close to the outer cylinder wall for various flow conditions.

Inspired by the previous measurement techniques and an understanding of the fluids flow another PIV campaign was set up to quantify the flow in the turbulent Taylor-Couette system. To resolve the large-scale circulation as well as the turbulent fluctuations a PIV at azimuthal-radial planes was performed and traversed along the axis of the system (Chapter 8). Thus, the PIV plane did scan through the whole volume and the three dimensional flow was resolved in the statistics of each layer while the camera captured the particle images though the transparent top plate of the experiment. From the temporal and spatial resolution the turbulent shear stress could be resolved partially. Of course smallest scales could not be resolved, but a comparison of the PIV and the direct torque measurements speaks for a well accomplished measurement. By the use of these measurements it is possible to distinguish how much the turbulent fluctuations contribute to the angular momentum transport and how much the large-scale circulation contributes. Special attention in this measurement was paid to the rotation ratio the torque is maximized, the pure inner cylinder rotation as well as the dependence in μ for a given shear Reynolds number.

By the use of all these measurement techniques, it is observed how turbulence and large-scale circulation are changed by the input parameters and influence the transport of angular momentum. Traditionally Taylor-Couette flow was investigated by rotating the inner cylinder and keeping the outer cylinder at rest ($\mu = 0$). For the flow visualisations also the laminar Taylor-vortices were seen for low Reynolds numbers. But the main interest of this study is the turbulent case, so all other measurements were taken for Reynolds numbers at a turbulent flow. For shear Reynolds numbers of about 10^3 up to $2 \cdot 10^4$ there are turbulent Taylor vortices observed for outer cylinder at rest. As the Reynolds number increases the strength of the large-scale circulation does decrease and the turbulence gains the upper hand at about $Re_S = 5 \cdot 10^4$. For $\Re_S = 8 \cdot 10^4$ and 10^5 it impossible to identify the large-scale circulation. In the radial profiles of the averaged radial velocity $\langle u_r \rangle_{A(r),t}$ it can be clearly seen that between $Re_S = 8 \cdot 10^4$ and $9 \cdot 10^4$ the radial component vanishes. For increasing driving velocity the flow behaviour remains like that. The large-scale circulation does not contribute to the global angular momentum flux, while the turbulent fluctuations take over the whole transport. Interestingly also the fluctuations in the boundary layers were increased at this shear Reynolds numbers and the ultimate regime is reached. The Nusselt number measured by torque meter agree with that, as there is a scaling change at $Re_S \approx 8 \cdot 10^4$. Afterwards the scaling approaches the value of $\beta = 0.76$, which is in agreement with other studies from Twente and Maryland for larger radius ratios done in the last years ([38], [30], [17], [31], [27]). As the Taylor number behaves as $Ta \sim Re_S^2$, the scaling of the Nusselt number with Taylor number is $Nu_\omega \sim Re_S^\beta \sim Ta^{\beta/2} = Ta^{0.38}$ the same as the Nusselt number of the heat flux with the Rayleigh number for Rayleigh-Bénard convection ($Nu_\Theta \approx Ra^{0.38}$ Grossmann and Lohse [14] and following publications). Thus, the analogy for the high turbulent flow can be confirmed.

As the pure inner cylinder rotation is traditionally considered as Taylor-Couette flow already Taylor [35] and Wendt [39] investigated the flow for both cylinders rotating. Later on only a few studies were performed to study these interesting dynamics. In Andereck et al. [1] the flow structures happening for co-rotating and counter-rotating cylinders at a radius ratio of $\eta = 0.86$ and shear Reynolds numbers below 5000 were explained. In the present thesis, the flow especially for counter rotating cylinders was analysed for shear Reynolds numbers starting at about 5000 and reaching larger than 10^6 at radius ratio $\eta = 0.5$.

For Re_S in the relative small turbulent regime ($Re_S < 5{\cdot}10^4$) the different flow structures are conducted. For co-rotating cylinders there is no observable feature in the flow. From pure inner cylinder rotation towards the maximum of torque the large-scale circulation is enhanced. For $Re_S < 2 \cdot 10^4$ the torque maximum still develops. The flow in this case was also studied by the DNS of Brauckmann and Eckhardt [2] and Ostilla-Mónico et al. [27] and compared with these experimental data. After the torque maximum the large-scale circulation detaches from the outer cylinder wall. The neutral surface detaches from the outer cylinder and the flow is stabilized by the outer linear stable flow. Inside the system still small vortices can be observed in the area of the inner instable regime. Brauckmann and Eckhardt [3] formulated the idea that an enlargement of the inner vortices can explain the position of the torque maximum by taking into account the growth of a vortex by Esser and Grossmann [12]. The prediction holds well with the observations made in the experiment and the flow structures are explained in detail in Chapters 5 to 8.

For large shear Reynolds numbers $Re_S > 10^5$ the behaviour of the flow structures is finally similar. The total estimated Nusselt number from the PIV is in good agreement with the direct torque measurements. So, both measurements can be well combined in the understanding of the flow. For pure inner cylinder rotation no more Taylor-vortices are identified, turbulent fluctuations take over the transport of angular momentum. As soon as a weak counter rotation sets on the large-scale circulation is brought up. In agreement with the prediction [3] for enhanced large-scale circulation the turbulent Taylor-vortices rise until the predicted torque maximum of $\mu = -0.20$. Coincidentally the turbulent fluctuations decrease with the increasing large-scale circulation. For the torque maximum also PIV measurements at shear Reynolds numbers up to $Re_S(\mu = -0.20) \leq 2.2 \cdot 10^5$ confirm the existence of the large-scale circulation causing the stronger angular momentum transport. For stronger counter rotation $\mu < -0.20$ the energy of the large-scale circulation drops and the turbulent fluctuations increase again. For the total budget the contributions of the large-scale circulation and the turbulence to the total Nusselt number are computed and confronted for the flow cases. In agreement with the flow velocity measurements the contribution of the large-scale circulation for pure inner cylinder rotation has no more effect onto the transport. For the slight counter rotation the large-scale circulation gains influence. At the case, that the large-scale circulation can grow into the gap most efficient [3] its contribution to the angular momentum transport becomes dominant. In the same time the turbulent transport has its minimum, but still the total transport is maximized. For the strong counter rotation the contribution of the large-scale circulation drops quickly and the turbulent fluctuations increase but remain stable.

Finally the turbulent Taylor-Couette flow for radius ratio $\eta = 0.5$ is widely investigated. The angular momentum transport in dependence on the input parameters (Re_S, μ) is revealed. The flow structures causing this dependencies are measured, analysed, explained and understood.

9.2 Outlook

The list of questions raised in front of this thesis has been successfully answered. In addition a few more questions have been risen during the study of the turbulent Taylor-Couette flow. The mainly interesting ones: What drives the torque to be maximal for counter rotation? Which flow structures dominate the counter rotation? have been also taken into account, investigated and explained. But the flow between two concentric cylinders is not yet fully understood.

Despite the measurements for this single geometry also the variation of the systems geometry needs to be studied. The main parameter for the geometry is the radius ratio. The TTCC experiment is relatively inflexible in terms of changing the dimensions. But the TvTCC experiment was intentionally built to give the ability of changing the inner cylinder. The system does have already the radius ratios $\eta = 0.1, 0.2, 0.29, 0.357, 0.5$ and 0.71 and in the meantime is equipped with a torque meter sensing also the torque for other geometries. The variation of the radius ratio is an ongoing process.

A time-resolved PIV would suit investigations of this system more and give a better understanding of coherent structures close to the wall. Due to the lack of availability a slow PIV system was used. Time-resolved PIV measurements are actually planned for counter-rotating flow at radius ratio $\eta = 0.2$. Also a parallel made measurement for time dependent flows as wavy vortices would be a good extension of the flow measurements. This would widen the understanding done in different heights at different times.

One observation during the experiments is the strong torques happening in the Rayleigh-stable regime. The underlying flow is of huge interest, as in this case no linear instability should happen. The huge torques are of interest for astrophysical flows as in accretion disks. There a source is searched explaining the observed large transport of angular momentum while objects as stars and planets are formed. The whole area of the quasi Keplerian flows $1 > \mu > \eta^2$ is still an open field to be studied in Taylor-Couette flow with large shear Reynolds numbers.

For the case of linear stable flows also the influence of the end walls becomes more dominant. For the present study the influence was checked for centrifugally unstable flows and showed less influence to the flow behaviour and torque measurements. Hence, its influence was neglected and can be an interesting area of research.

From technical point of view, the both experiments can be used to study drag reduction in rotating systems. On one hand the torque sensing gives a direct measure if a reduction method works or not and on the other hand the understanding of the fluid mechanics gained can be used to set up drag reduction methods. For example, now is clear that for outer cylinder at rest, highly rotating systems do not show up the large-scale circulation. Thus, trying to prevent these structures does not play any role.

Another interesting topic would be to study the influence of larger particles in the Taylor-Couette system. The effect of particle banding, where particles of similar size are transported to a certain region in the flow, a band. Thus, the system could be used for the classification of particles in co-rotating regime. It is also of interest, that for the torque maximum, the mixing of the system is huge while the turbulent shear stresses are reduced. For processes, where low shear stresses but efficient mixing is required the counter-rotating Taylor-Couette flow can be an ideal reactor. One example would be chemical engineering, where biological samples are needed. Another would be the flocculation of material, as this can happen more efficient in the large-scale circulation with low turbulent stresses but allows larger passage of used fluids.

Despite these few ideas, it remains a huge number of influences and variations, which the Taylor-Couette system still keeps as its secrets.

Appendix A

Appendix A: Flow Visualisation

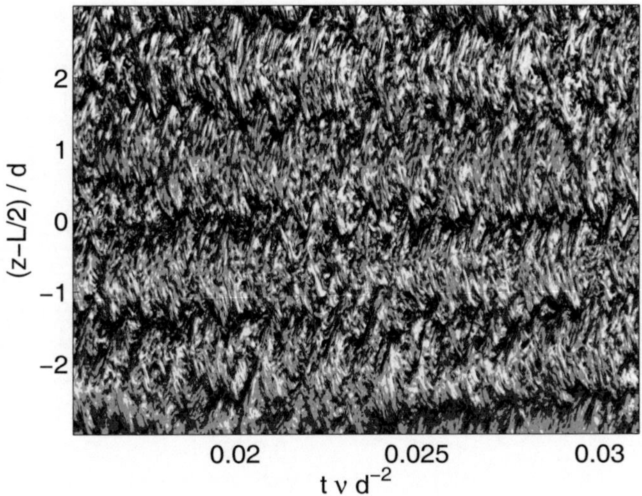

FIGURE A.1: Space time diagram of slight counter rotating turbulent Taylor-Couette flow for $Re_S = 52000$ and $\mu = 0.00$. Second fragment of measured signal depicted in Figure 6.14.

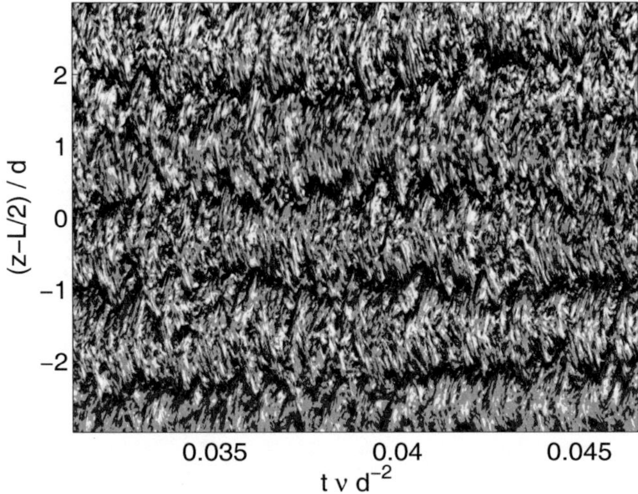

FIGURE A.2: Space time diagram of slight counter rotating turbulent Taylor-Couette flow for $Re_S = 52000$ and $\mu = 0.00$. Third fragment of measured signal depicted in Figure 6.14.

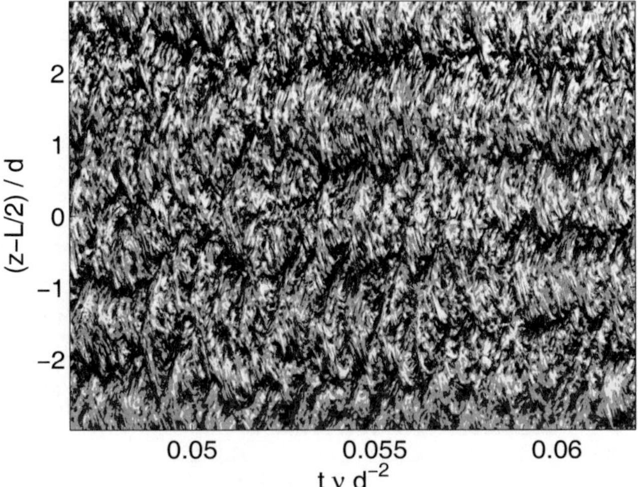

FIGURE A.3: Space time diagram of slight counter rotating turbulent Taylor-Couette flow for $Re_S = 52000$ and $\mu = 0.00$. Fourth fragment of measured signal depicted in Figure 6.14.

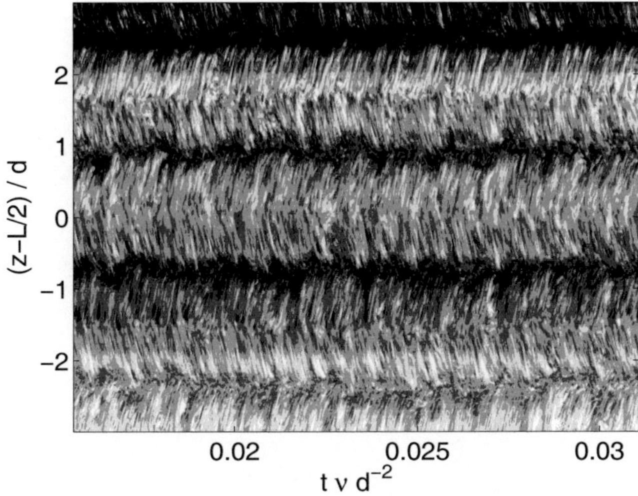

FIGURE A.4: Space time diagram of slight counter rotating turbulent Taylor-Couette flow for $Re_S = 52000$ and $\mu = -0.10$. Second fragment of measured signal depicted in Figure 6.15.

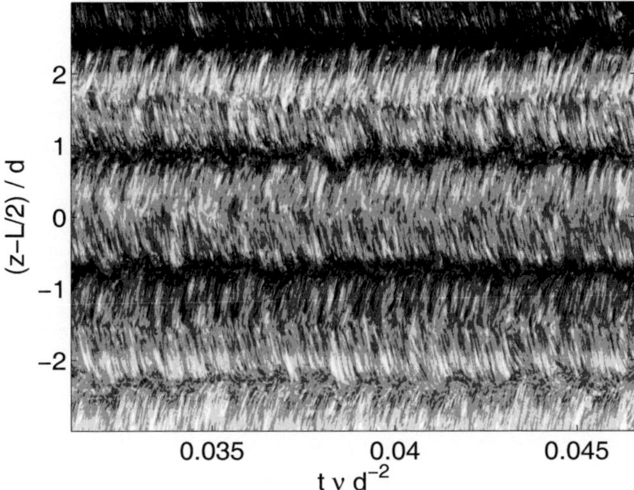

FIGURE A.5: Space time diagram of slight counter rotating turbulent Taylor-Couette flow for $Re_S = 52000$ and $\mu = -0.10$. Third fragment of measured signal depicted in Figure 6.15.

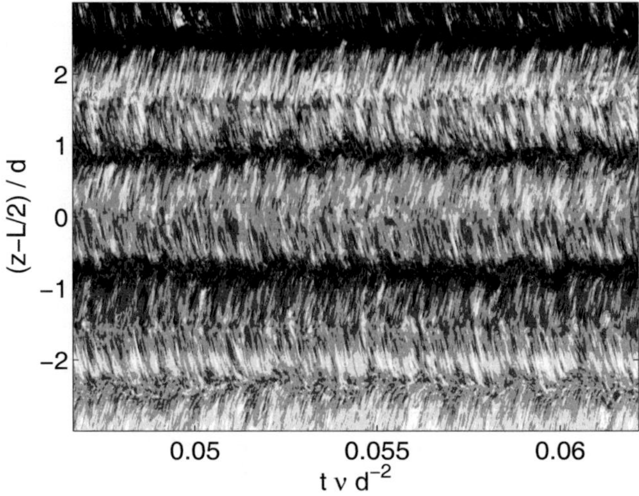

FIGURE A.6: Space time diagram of slight counter rotating turbulent Taylor-Couette flow for $Re_S = 52000$ and $\mu = -0.10$. Fourth fragment of measured signal depicted in Figure 6.16.

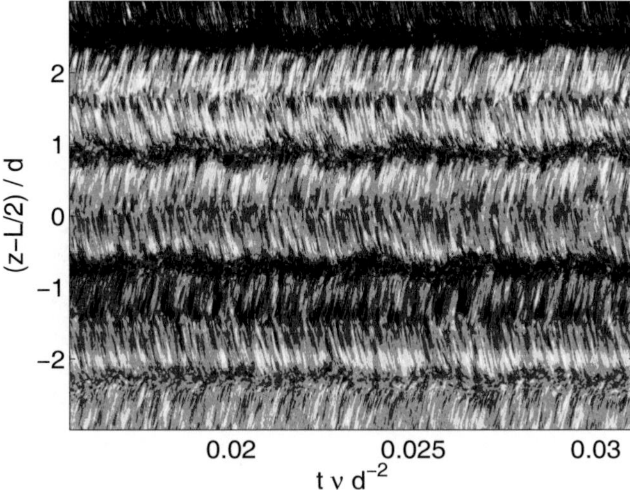

FIGURE A.7: Space time diagram of slight counter rotating turbulent Taylor-Couette flow for $Re_S = 52000$ and $\mu = -0.15$. Second fragment of measured signal depicted in Figure 6.16.

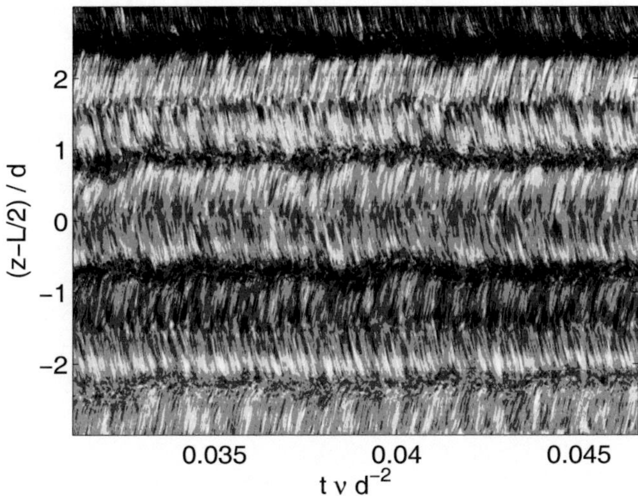

FIGURE A.8: Space time diagram of slight counter rotating turbulent Taylor-Couette flow for $Re_S = 52000$ and $\mu = -0.15$. Third fragment of measured signal depicted in Figure 6.16.

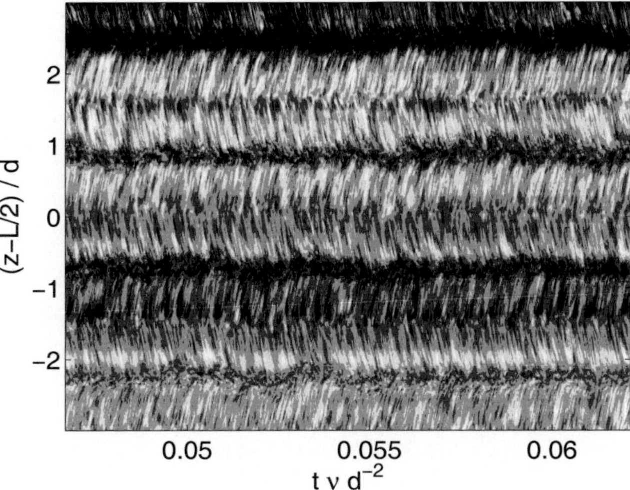

FIGURE A.9: Space time diagram of slight counter rotating turbulent Taylor-Couette flow for $Re_S = 52000$ and $\mu = -0.15$. Fourth fragment of measured signal depicted in Figure 6.16.

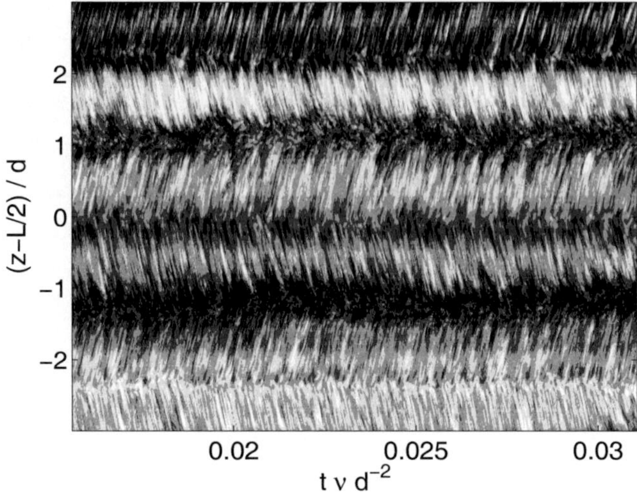

FIGURE A.10: Space time diagram of slight counter rotating turbulent Taylor-Couette flow for $Re_S = 52000$ and $\mu = -0.20$. Second fragment of measured signal depicted in Figure 6.17.

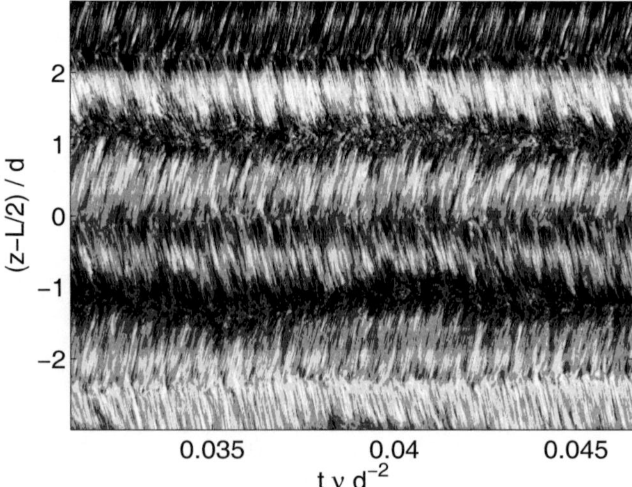

FIGURE A.11: Space time diagram of slight counter rotating turbulent Taylor-Couette flow for $Re_S = 52000$ and $\mu = -0.20$. Third fragment of measured signal depicted in Figure 6.17.

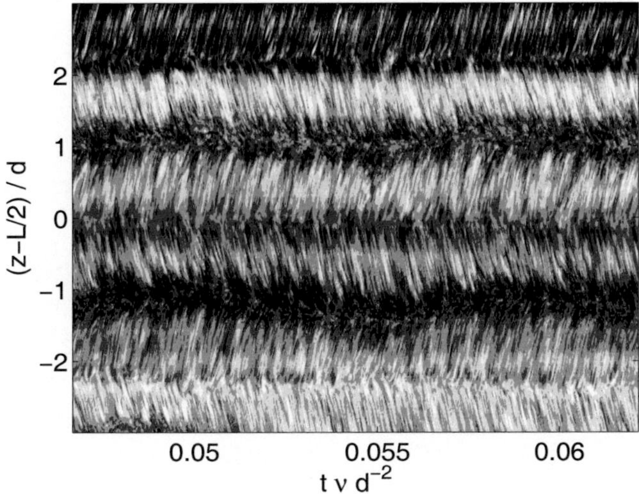

FIGURE A.12: Space time diagram of slight counter rotating turbulent Taylor-Couette flow for $Re_S = 52000$ and $\mu = -0.20$. Fourth fragment of measured signal depicted in Figure 6.17.

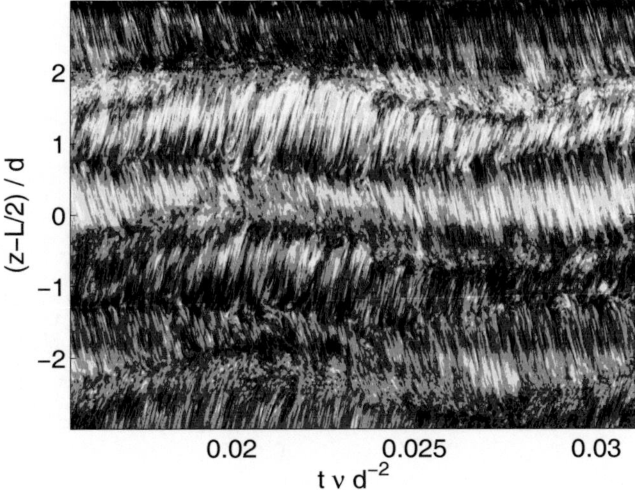

FIGURE A.13: Space time diagram of slight counter rotating turbulent Taylor-Couette flow for $Re_S = 52000$ and $\mu = -0.25$. Second fragment of measured signal depicted in Figure 6.18.

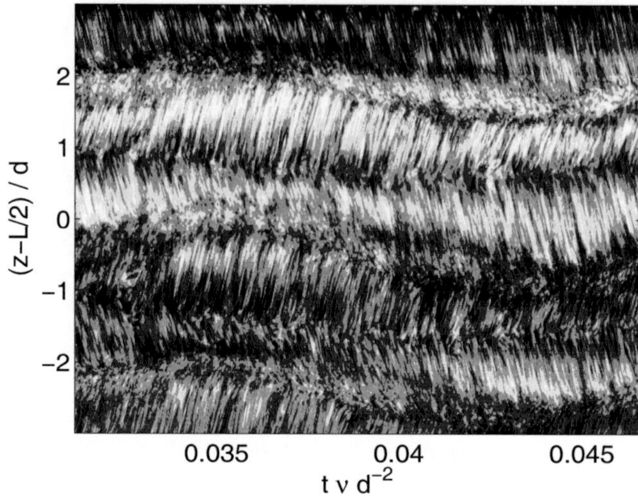

FIGURE A.14: Space time diagram of slight counter rotating turbulent Taylor-Couette flow for $Re_S = 52000$ and $\mu = -0.25$. Third fragment of measured signal depicted in Figure 6.18.

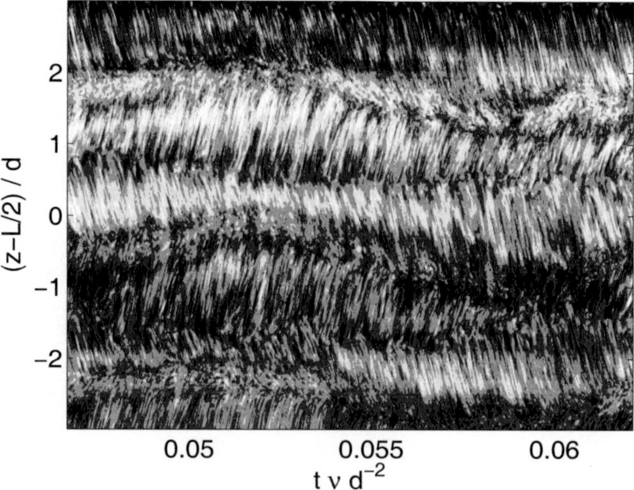

FIGURE A.15: Space time diagram of slight counter rotating turbulent Taylor-Couette flow for $Re_S = 52000$ and $\mu = -0.25$. Fourth fragment of measured signal depicted in Figure 6.18.

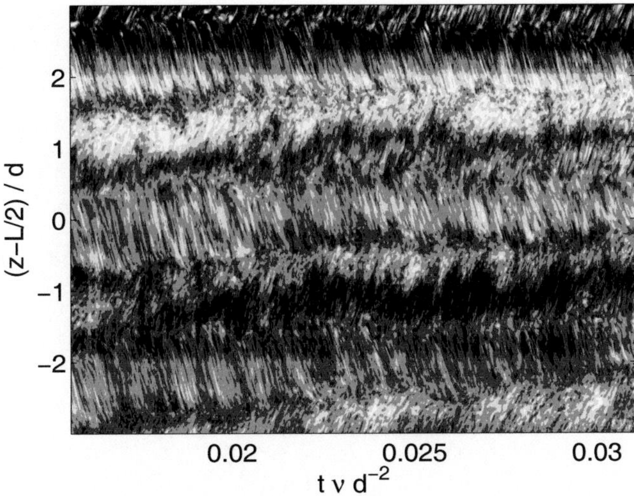

FIGURE A.16: Space time diagram of slight counter rotating turbulent Taylor-Couette flow for $Re_S = 52000$ and $\mu = -0.30$. Second fragment of measured signal depicted in Figure 6.19.

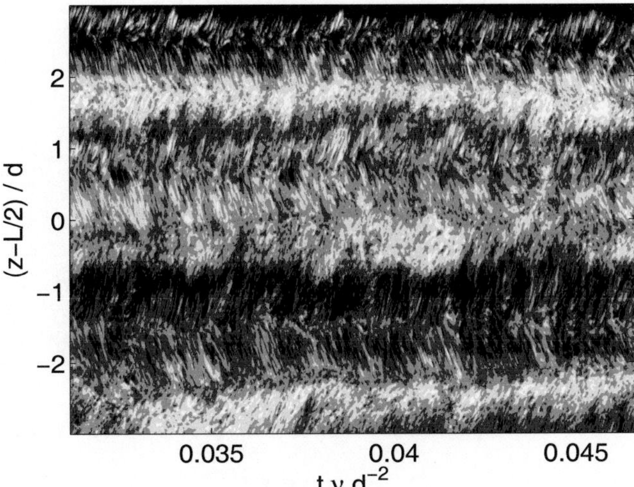

FIGURE A.17: Space time diagram of slight counter rotating turbulent Taylor-Couette flow for $Re_S = 52000$ and $\mu = -0.30$. Third fragment of measured signal depicted in Figure 6.19.

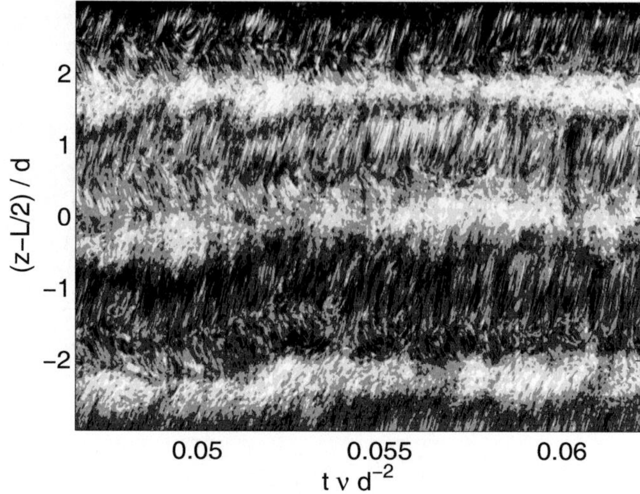

FIGURE A.18: Space time diagram of slight counter rotating turbulent Taylor-Couette flow for $Re_S = 52000$ and $\mu = -0.30$. Fourth fragment of measured signal depicted in Figure 6.19.

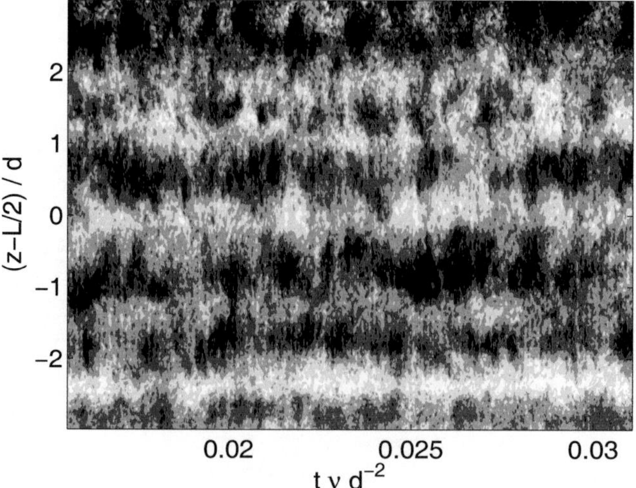

FIGURE A.19: Space time diagram of slight counter rotating turbulent Taylor-Couette flow for $Re_S = 52000$ and $\mu = -0.50$. Second fragment of measured signal depicted in Figure 6.20.

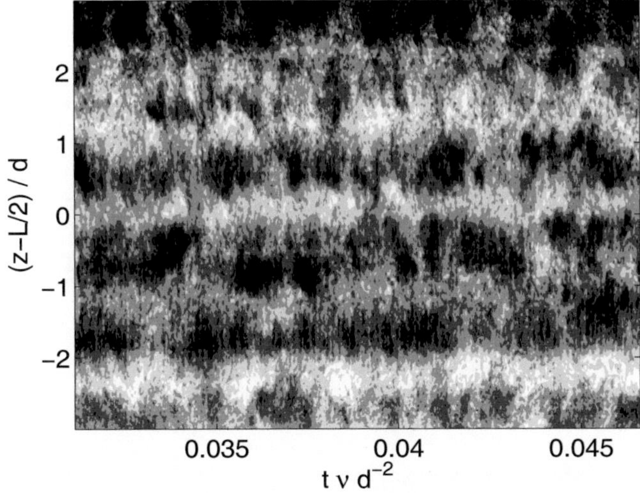

FIGURE A.20: Space time diagram of slight counter rotating turbulent Taylor-Couette flow for $Re_S = 52000$ and $\mu = -0.50$. Third fragment of measured signal depicted in Figure 6.20.

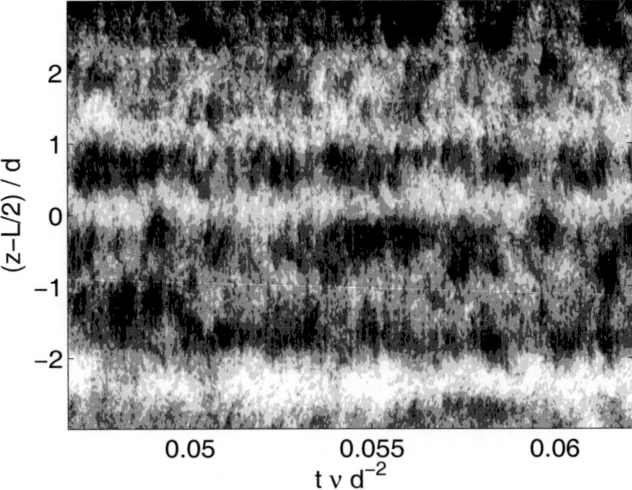

FIGURE A.21: Space time diagram of slight counter rotating turbulent Taylor-Couette flow for $Re_S = 52000$ and $\mu = -0.50$. Fourth fragment of measured signal depicted in Figure 6.20.

Appendix B

Appendix B: Particle Image Velocimetry

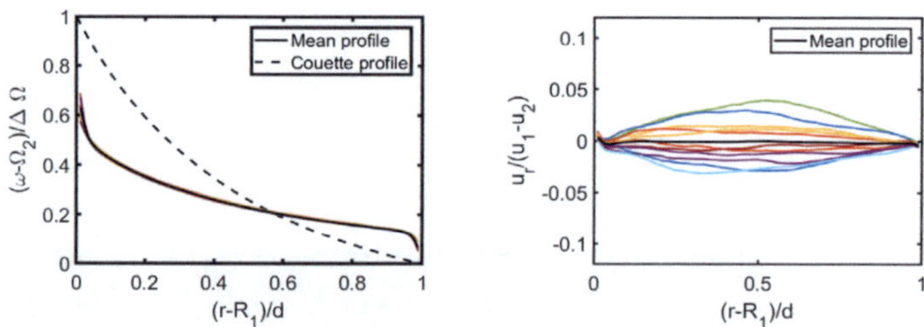

FIGURE B.1: Angular and radial velocity profiles averaged in time and azimuthal coordinate for $Re_S = 40000$, $\mu = 0$. In colour the different axial positions are depicted, the black solid line represents the axial mean profile and the dashed line in angular velocity profile represents the laminar Couette-solution.

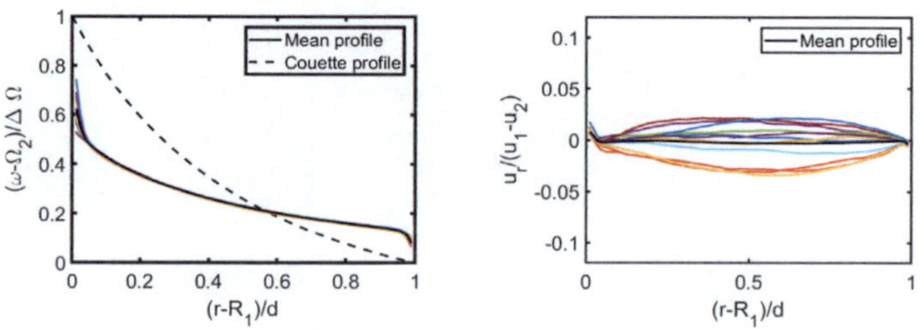

FIGURE B.2: Angular and radial velocity profiles averaged in time and azimuthal coordinate for $Re_S = 50000$, $\mu = 0$. In colour the different axial positions are depicted, the black solid line represents the axial mean profile and the dashed line in angular velocity profile represents the laminar Couette-solution.

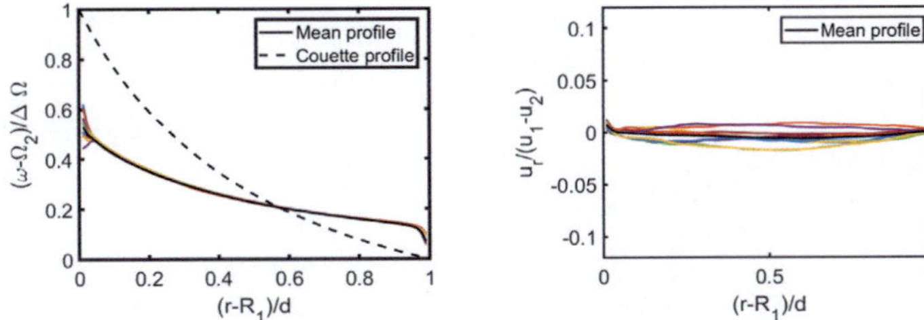

FIGURE B.3: Angular and radial velocity profiles averaged in time and azimuthal coordinate for $Re_S = 60000$, $\mu = 0$. In colour the different axial positions are depicted, the black solid line represents the axial mean profile and the dashed line in angular velocity profile represents the laminar Couette-solution.

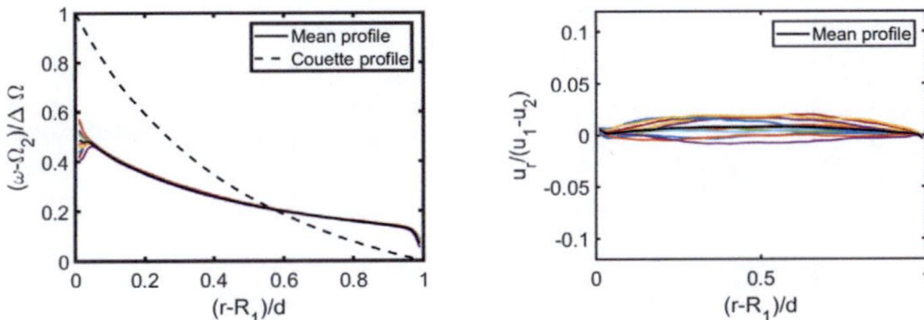

FIGURE B.4: Angular and radial velocity profiles averaged in time and azimuthal coordinate for $Re_S = 70000$, $\mu = 0$. In colour the different axial positions are depicted, the black solid line represents the axial mean profile and the dashed line in angular velocity profile represents the laminar Couette-solution.

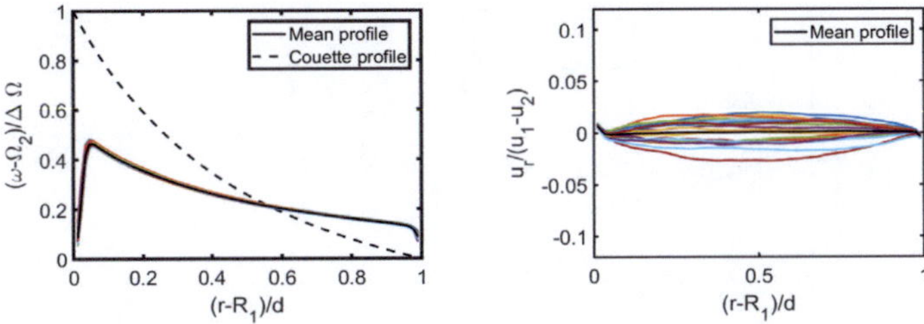

FIGURE B.5: Angular and radial velocity profiles averaged in time and azimuthal coordinate for $Re_S = 80000$, $\mu = 0$. In colour the different axial positions are depicted, the black solid line represents the axial mean profile and the dashed line in angular velocity profile represents the laminar Couette-solution.

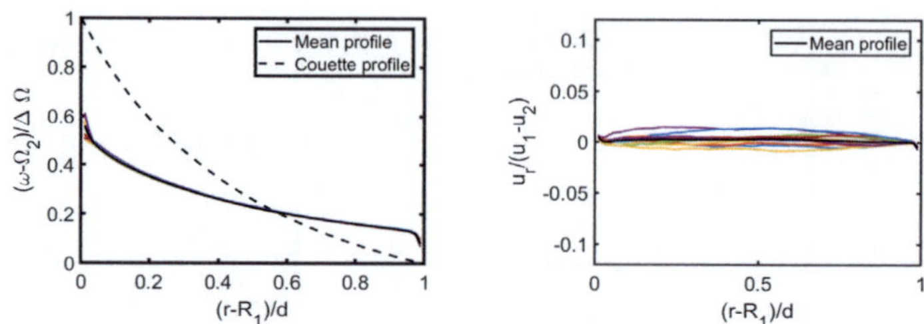

FIGURE B.6: Angular and radial velocity profiles averaged in time and azimuthal coordinate for $Re_S = 90000$, $\mu = 0$. In colour the different axial positions are depicted, the black solid line represents the axial mean profile and the dashed line in angular velocity profile represents the laminar Couette-solution.

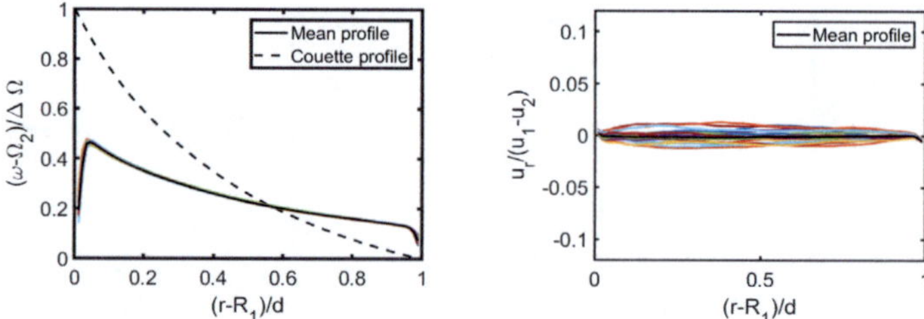

FIGURE B.7: Angular and radial velocity profiles averaged in time and azimuthal coordinate for $Re_S = 100000$, $\mu = 0$. In colour the different axial positions are depicted, the black solid line represents the axial mean profile and the dashed line in angular velocity profile represents the laminar Couette-solution.

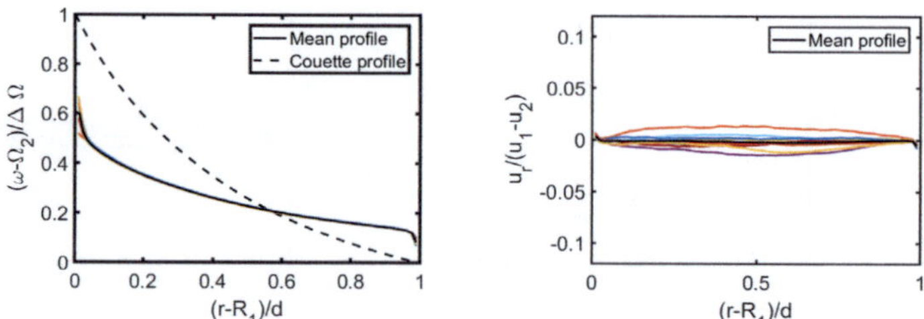

FIGURE B.8: Angular and radial velocity profiles averaged in time and azimuthal coordinate for $Re_S = 110000$, $\mu = 0$. In colour the different axial positions are depicted, the black solid line represents the axial mean profile and the dashed line in angular velocity profile represents the laminar Couette-solution.

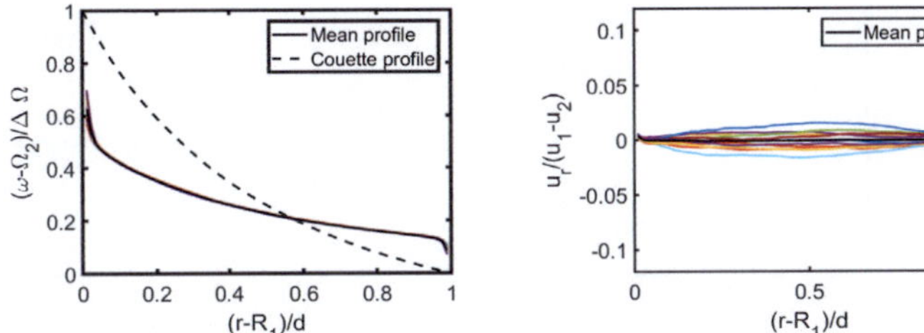

FIGURE B.9: Angular and radial velocity profiles averaged in time and azimuthal coordinate for $Re_S = 120000$, $\mu = 0$. In colour the different axial positions are depicted, the black solid line represents the axial mean profile and the dashed line in angular velocity profile represents the laminar Couette-solution.

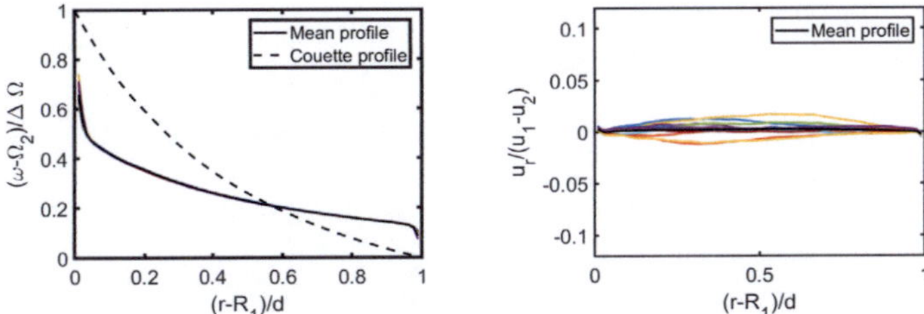

FIGURE B.10: Angular and radial velocity profiles averaged in time and azimuthal coordinate for $Re_S = 130000$, $\mu = 0$. In colour the different axial positions are depicted, the black solid line represents the axial mean profile and the dashed line in angular velocity profile represents the laminar Couette-solution.

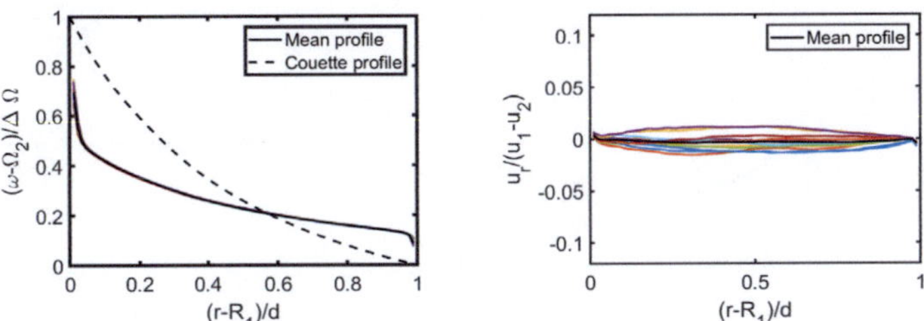

FIGURE B.11: Angular and radial velocity profiles averaged in time and azimuthal coordinate for $Re_S = 140000$, $\mu = 0$. In colour the different axial positions are depicted, the black solid line represents the axial mean profile and the dashed line in angular velocity profile represents the laminar Couette-solution.

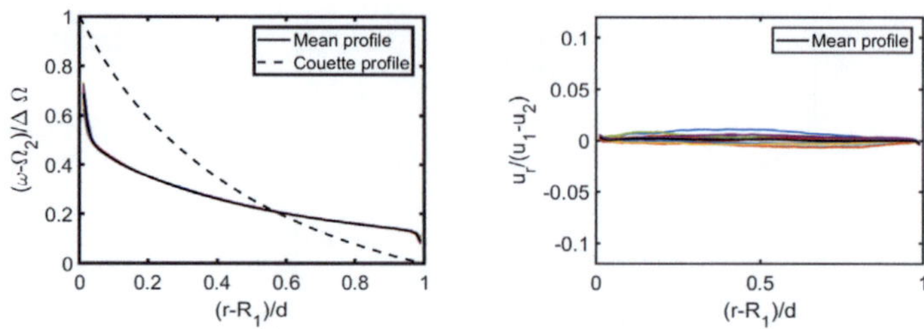

FIGURE B.12: Angular and radial velocity profiles averaged in time and azimuthal coordinate for $Re_S = 150000$, $\mu = 0$. In colour the different axial positions are depicted, the black solid line represents the axial mean profile and the dashed line in angular velocity profile represents the laminar Couette-solution.

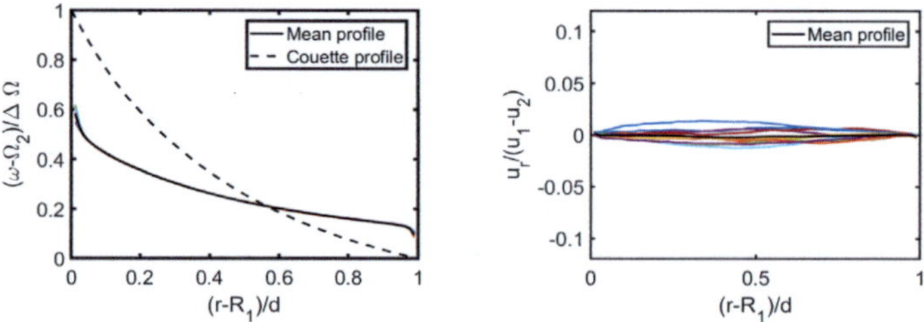

FIGURE B.13: Angular and radial velocity profiles averaged in time and azimuthal coordinate for $Re_S = 165000$, $\mu = 0$. In colour the different axial positions are depicted, the black solid line represents the axial mean profile and the dashed line in angular velocity profile represents the laminar Couette-solution.

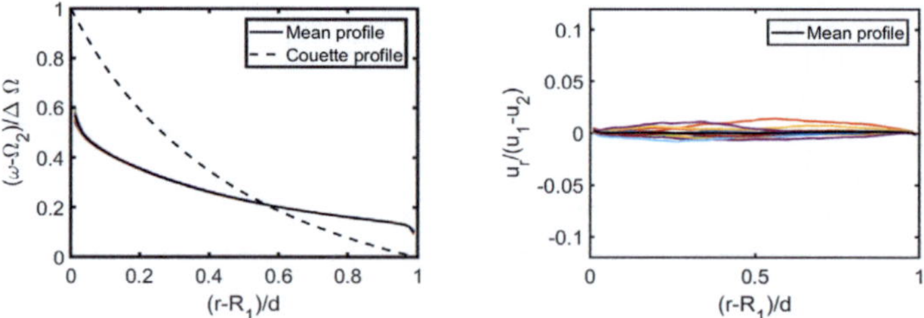

FIGURE B.14: Angular and radial velocity profiles averaged in time and azimuthal coordinate for $Re_S = 192000$, $\mu = 0$. In colour the different axial positions are depicted, the black solid line represents the axial mean profile and the dashed line in angular velocity profile represents the laminar Couette-solution.

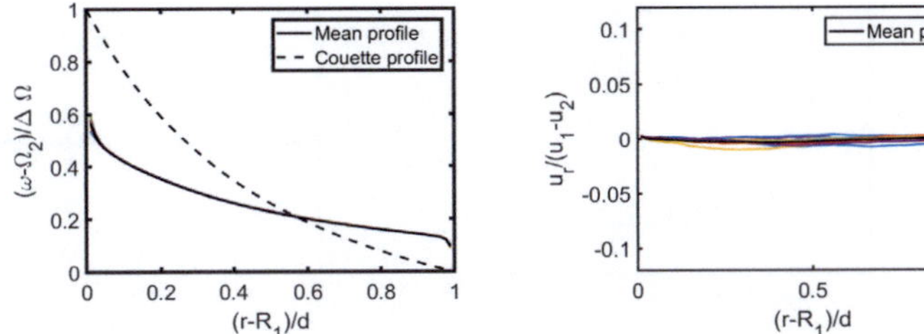

FIGURE B.15: Angular and radial velocity profiles averaged in time and azimuthal coordinate for $Re_S = 201000$, $\mu = 0$. In colour the different axial positions are depicted, the black solid line represents the axial mean profile and the dashed line in angular velocity profile represents the laminar Couette-solution.

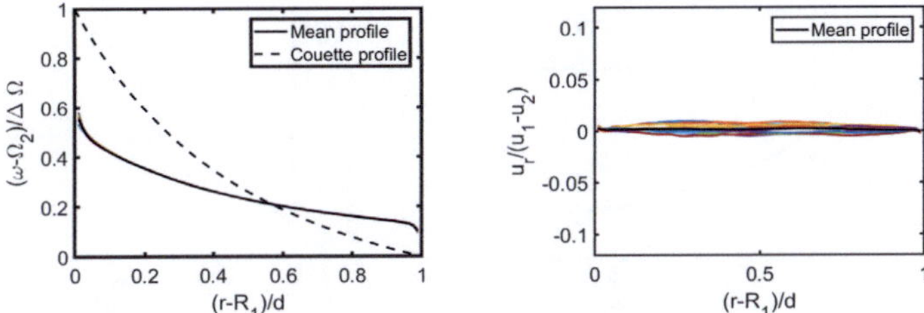

FIGURE B.16: Angular and radial velocity profiles averaged in time and azimuthal coordinate for $Re_S = 219000$, $\mu = 0$. In colour the different axial positions are depicted, the black solid line represents the axial mean profile and the dashed line in angular velocity profile represents the laminar Couette-solution.

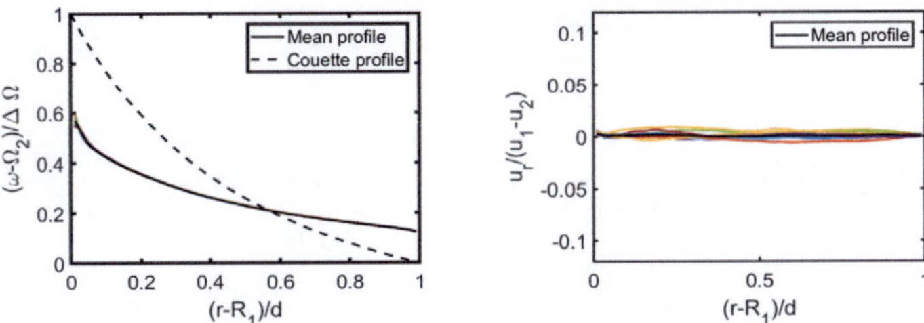

FIGURE B.17: Angular and radial velocity profiles averaged in time and azimuthal coordinate for $Re_S = 245000$, $\mu = 0$. In colour the different axial positions are depicted, the black solid line represents the axial mean profile and the dashed line in angular velocity profile represents the laminar Couette-solution.

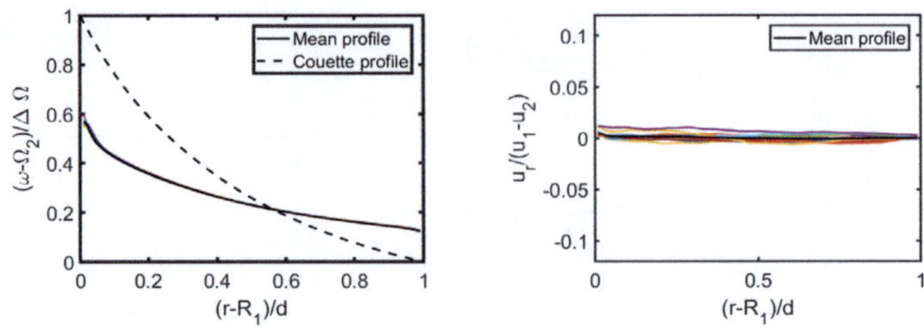

FIGURE B.18: Angular and radial velocity profiles averaged in time and azimuthal coordinate for $Re_S = 258000$, $\mu = 0$. In colour the different axial positions are depicted, the black solid line represents the axial mean profile and the dashed line in angular velocity profile represents the laminar Couette-solution.

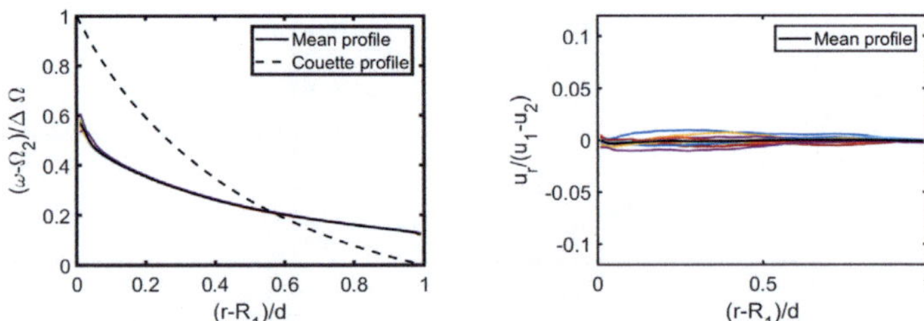

FIGURE B.19: Angular and radial velocity profiles averaged in time and azimuthal coordinate for $Re_S = 270000$, $\mu = 0$. In colour the different axial positions are depicted, the black solid line represents the axial mean profile and the dashed line in angular velocity profile represents the laminar Couette-solution.

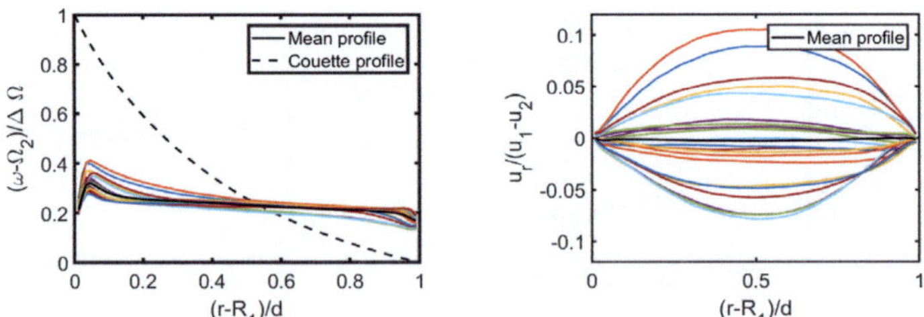

FIGURE B.20: Angular and radial velocity profiles averaged in time and azimuthal coordinate for $Re_S = 80000$, $\mu = -0.20$. In colour the different axial positions are depicted, the black solid line represents the axial mean profile and the dashed line in angular velocity profile represents the laminar Couette-solution.

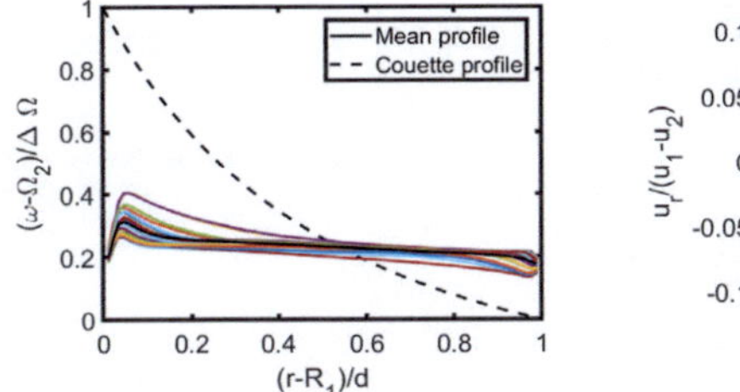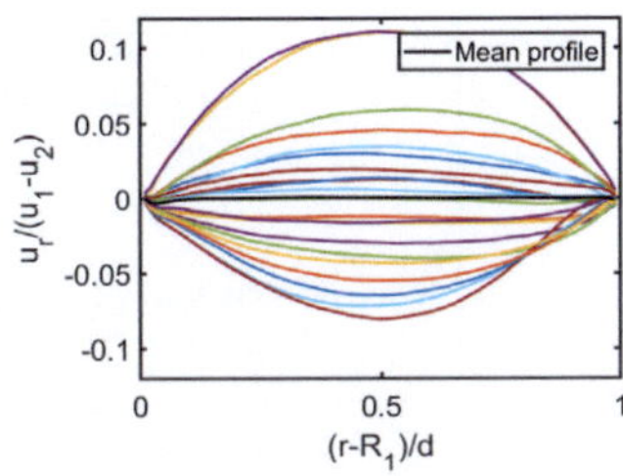

FIGURE B.21: Angular and radial velocity profiles averaged in time and azimuthal coordinate for $Re_S = 100000$, $\mu = -0.20$. In colour the different axial positions are depicted, the black solid line represents the axial mean profile and the dashed line in angular velocity profile represents the laminar Couette-solution.

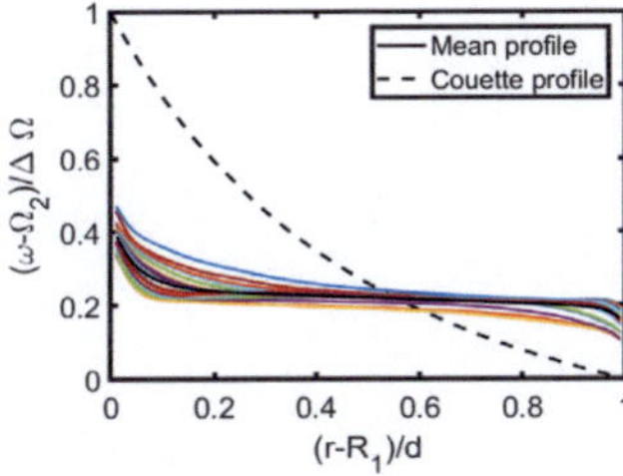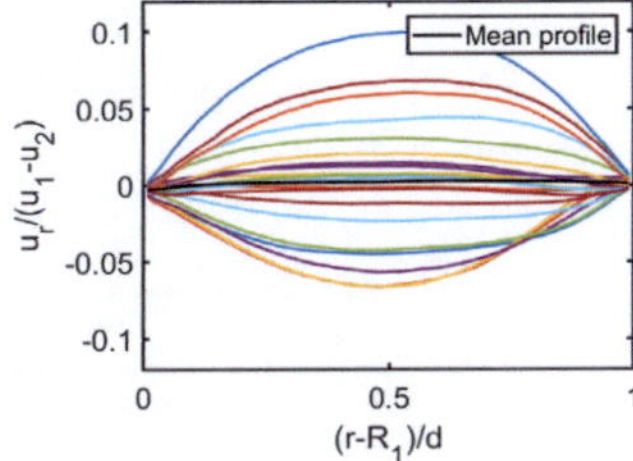

FIGURE B.22: Angular and radial velocity profiles averaged in time and azimuthal coordinate for $Re_S = 165000$, $\mu = -0.20$. In colour the different axial positions are depicted, the black solid line represents the axial mean profile and the dashed line in angular velocity profile represents the laminar Couette-solution.

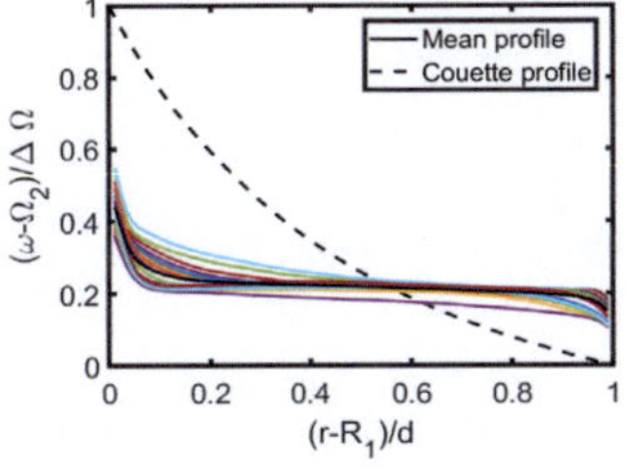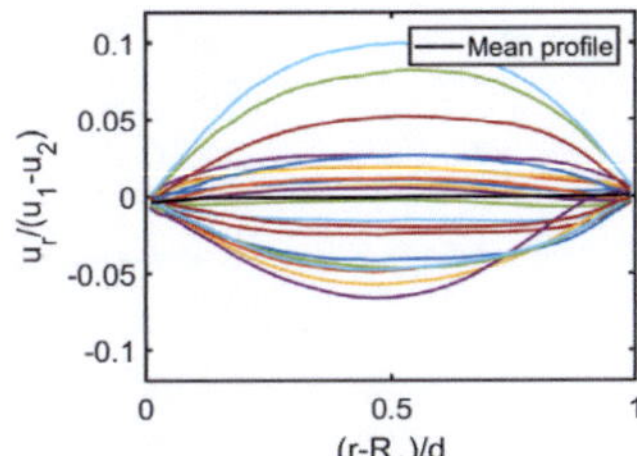

FIGURE B.23: Angular and radial velocity profiles averaged in time and azimuthal coordinate for $Re_S = 192000$, $\mu = -0.20$. In colour the different axial positions are depicted, the black solid line represents the axial mean profile and the dashed line in angular velocity profile represents the laminar Couette-solution.

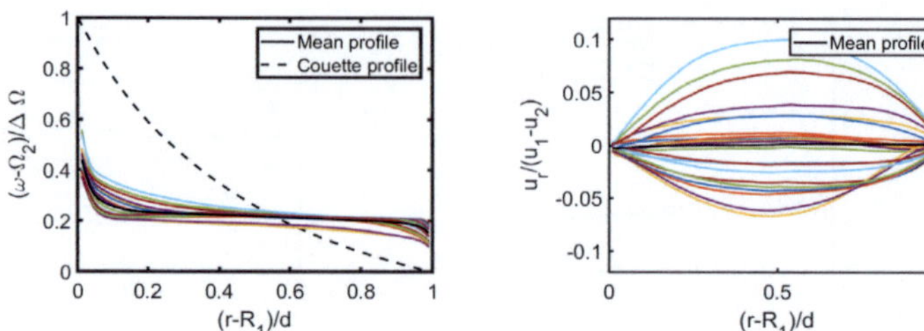

FIGURE B.24: Angular and radial velocity profiles averaged in time and azimuthal coordinate for $Re_S = 201000$, $\mu = -0.20$. In colour the different axial positions are depicted, the black solid line represents the axial mean profile and the dashed line in angular velocity profile represents the laminar Couette-solution.

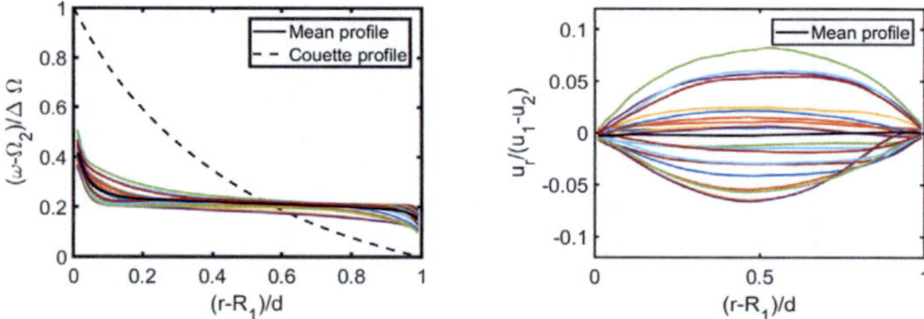

FIGURE B.25: Angular and radial velocity profiles averaged in time and azimuthal coordinate for $Re_S = 219000$, $\mu = -0.20$. In colour the different axial positions are depicted, the black solid line represents the axial mean profile and the dashed line in angular velocity profile represents the laminar Couette-solution.

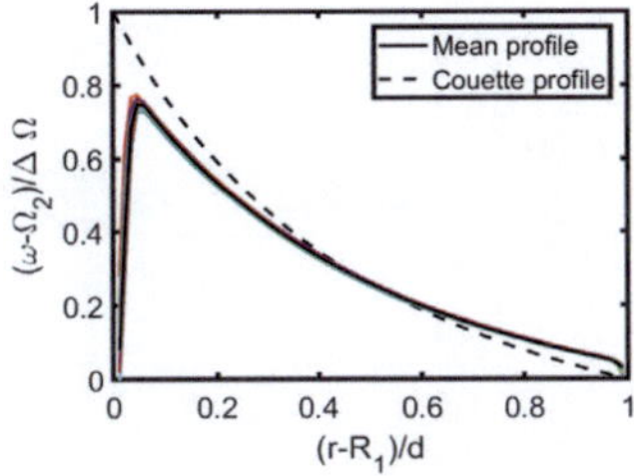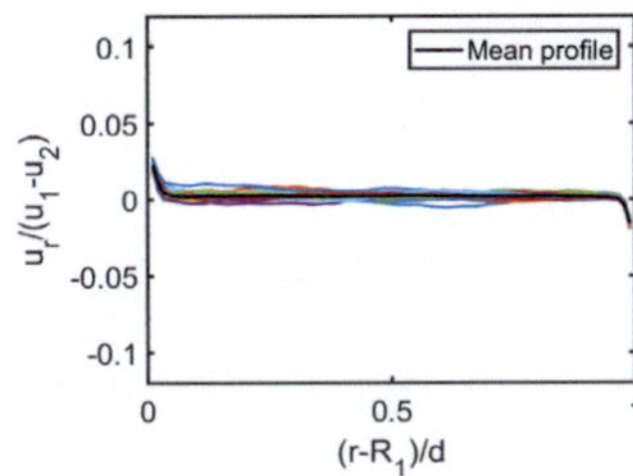

FIGURE B.26: Angular and radial velocity profiles averaged in time and azimuthal coordinate for $Re_S = 80000$, $\mu = +0.20$. In colour the different axial positions are depicted, the black solid line represents the axial mean profile and the dashed line in angular velocity profile represents the laminar Couette-solution.

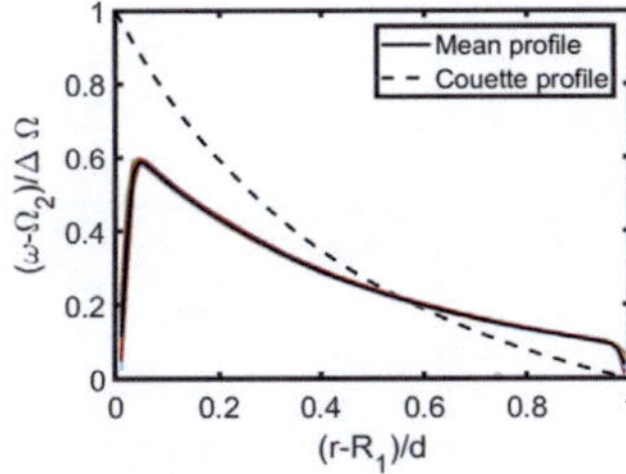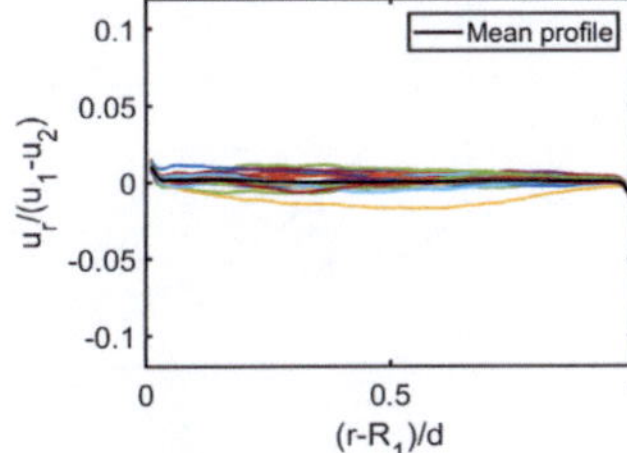

FIGURE B.27: Angular and radial velocity profiles averaged in time and azimuthal coordinate for $Re_S = 80000$, $\mu = +0.10$. In colour the different axial positions are depicted, the black solid line represents the axial mean profile and the dashed line in angular velocity profile represents the laminar Couette-solution.

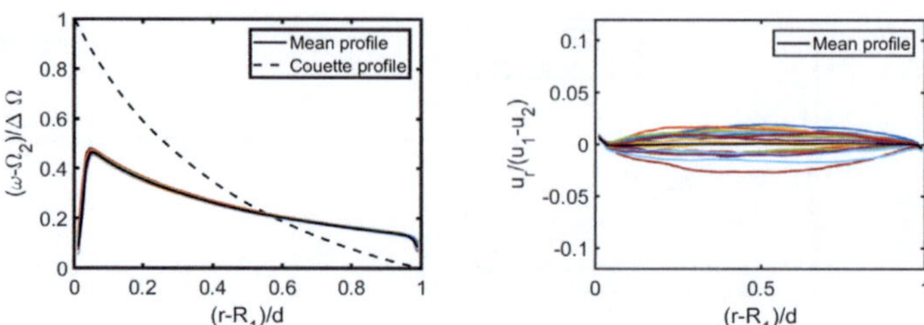

FIGURE B.28: Angular and radial velocity profiles averaged in time and azimuthal coordinate for $Re_S = 80000$, $\mu = 0$. In colour the different axial positions are depicted, the black solid line represents the axial mean profile and the dashed line in angular velocity profile represents the laminar Couette-solution.

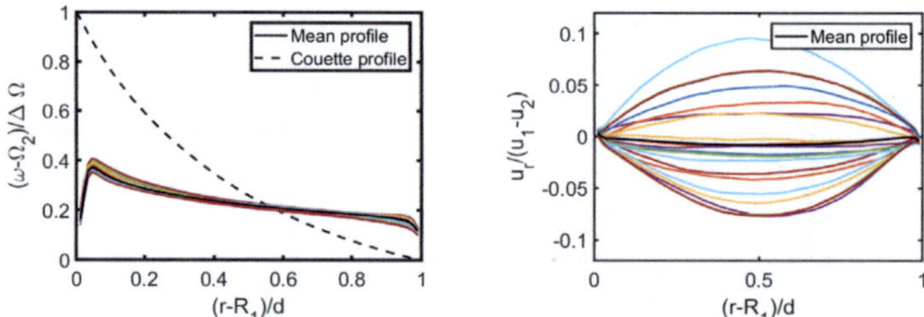

FIGURE B.29: Angular and radial velocity profiles averaged in time and azimuthal coordinate for $Re_S = 80000$, $\mu = -0.10$. In colour the different axial positions are depicted, the black solid line represents the axial mean profile and the dashed line in angular velocity profile represents the laminar Couette-solution.

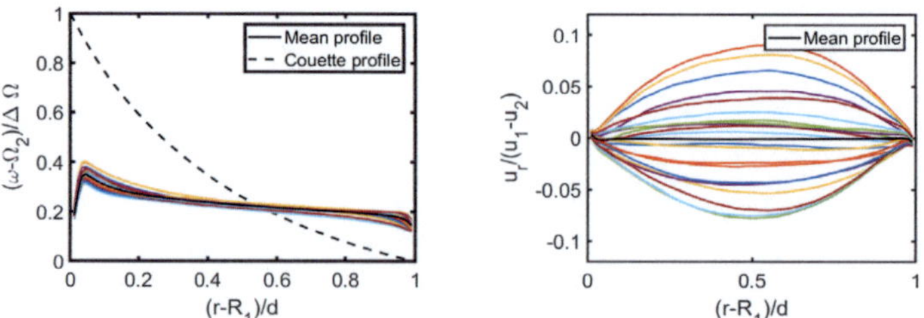

FIGURE B.30: Angular and radial velocity profiles averaged in time and azimuthal coordinate for $Re_S = 80000$, $\mu = -0.15$. In colour the different axial positions are depicted, the black solid line represents the axial mean profile and the dashed line in angular velocity profile represents the laminar Couette-solution.

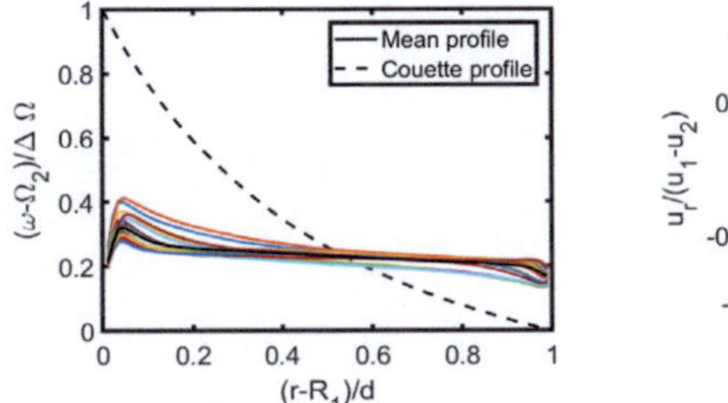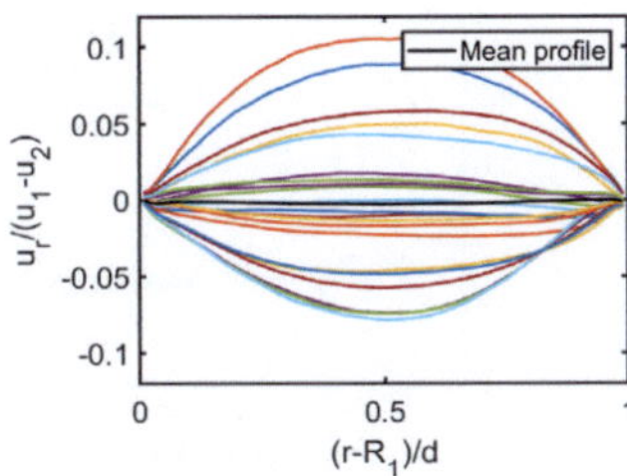

FIGURE B.31: Angular and radial velocity profiles averaged in time and azimuthal coordinate for $Re_S = 80000$, $\mu = -0.20$. In colour the different axial positions are depicted, the black solid line represents the axial mean profile and the dashed line in angular velocity profile represents the laminar Couette-solution.

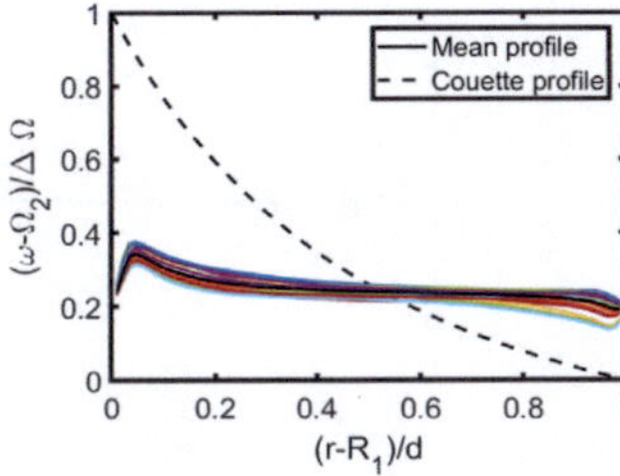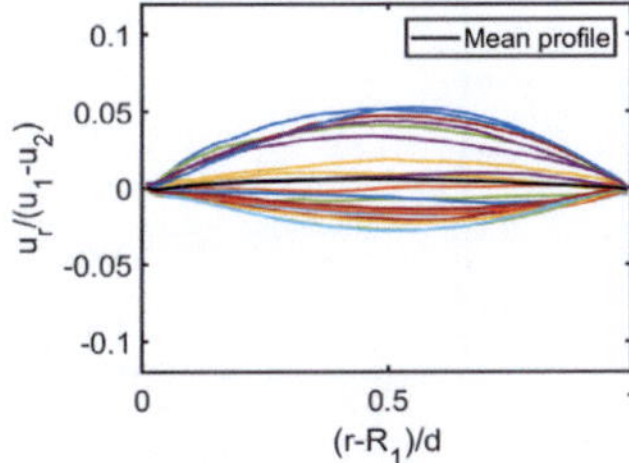

FIGURE B.32: Angular and radial velocity profiles averaged in time and azimuthal coordinate for $Re_S = 80000$, $\mu = -0.25$. In colour the different axial positions are depicted, the black solid line represents the axial mean profile and the dashed line in angular velocity profile represents the laminar Couette-solution.

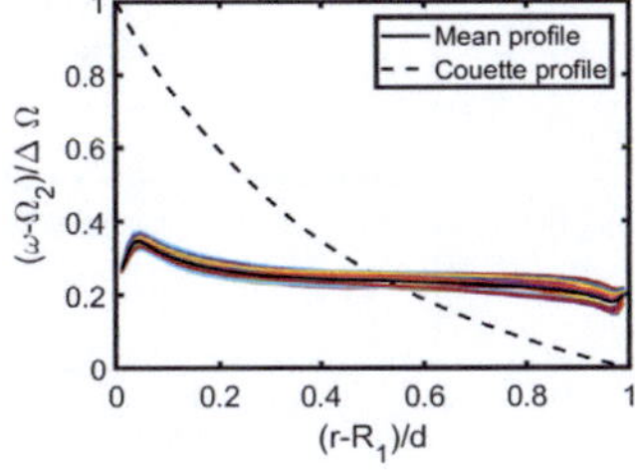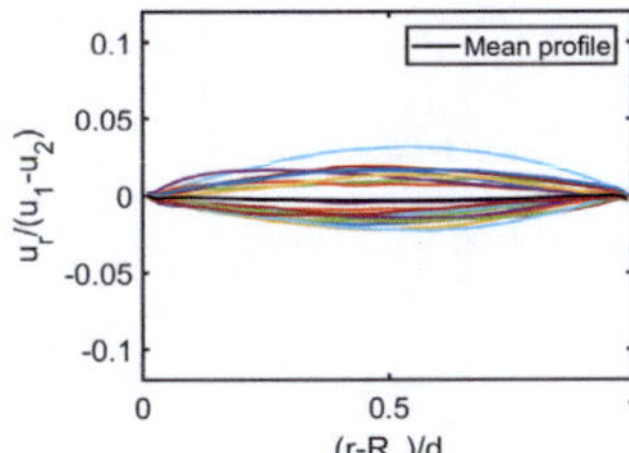

FIGURE B.33: Angular and radial velocity profiles averaged in time and azimuthal coordinate for $Re_S = 80000$, $\mu = -0.30$. In colour the different axial positions are depicted, the black solid line represents the axial mean profile and the dashed line in angular velocity profile represents the laminar Couette-solution.

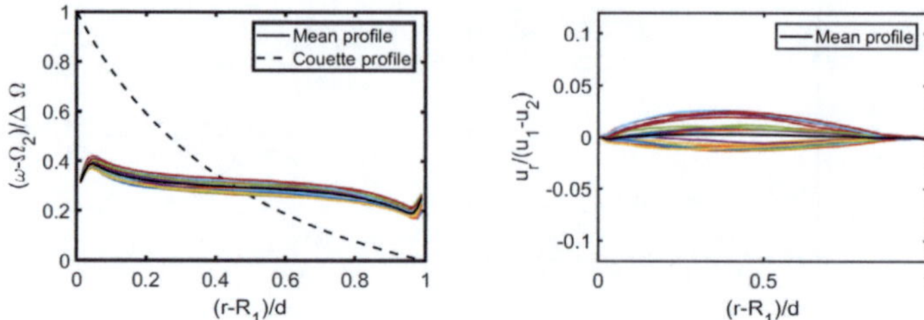

FIGURE B.34: Angular and radial velocity profiles averaged in time and azimuthal coordinate for $Re_S = 80000$, $\mu = -0.40$. In colour the different axial positions are depicted, the black solid line represents the axial mean profile and the dashed line in angular velocity profile represents the laminar Couette-solution.

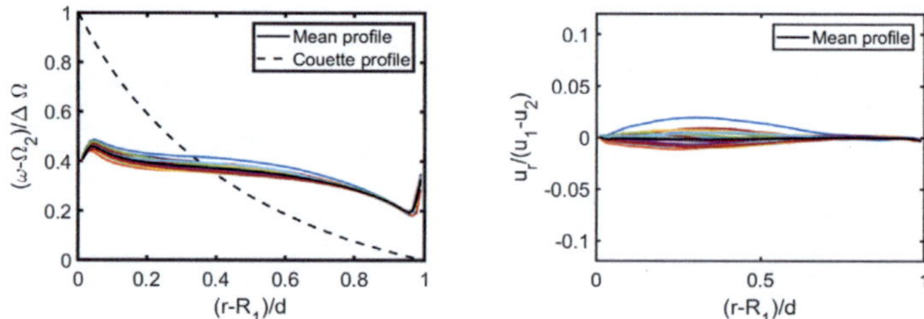

FIGURE B.35: Angular and radial velocity profiles averaged in time and azimuthal coordinate for $Re_S = 80000$, $\mu = -0.60$. In colour the different axial positions are depicted, the black solid line represents the axial mean profile and the dashed line in angular velocity profile represents the laminar Couette-solution.

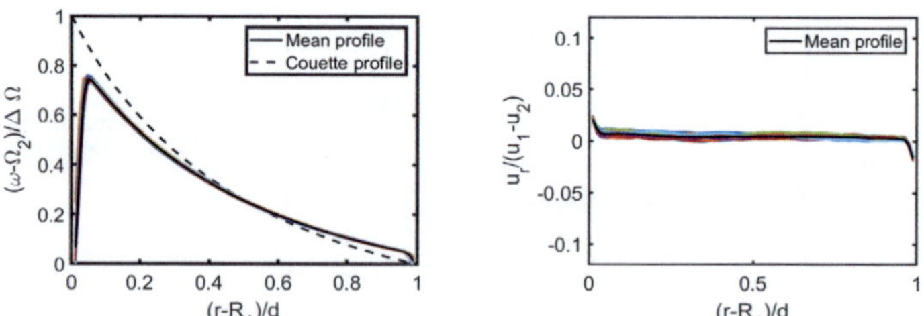

FIGURE B.36: Angular and radial velocity profiles averaged in time and azimuthal coordinate for $Re_S = 100000$, $\mu = +0.20$. In colour the different axial positions are depicted, the black solid line represents the axial mean profile and the dashed line in angular velocity profile represents the laminar Couette-solution.

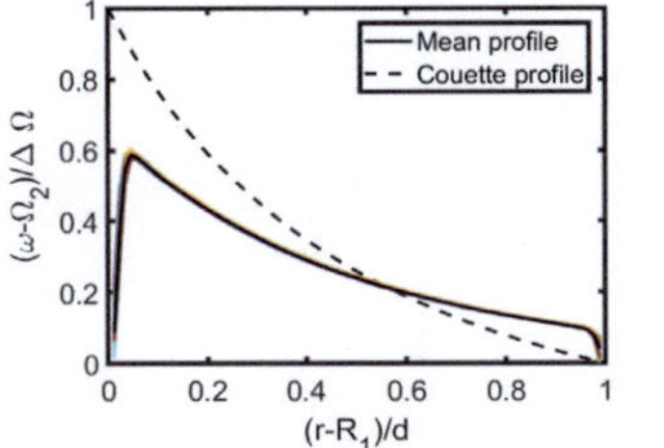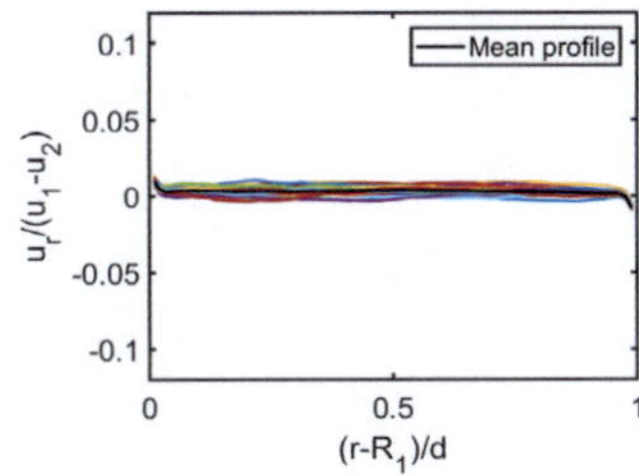

FIGURE B.37: Angular and radial velocity profiles averaged in time and azimuthal coordinate for $Re_S = 100000$, $\mu = +0.10$. In colour the different axial positions are depicted, the black solid line represents the axial mean profile and the dashed line in angular velocity profile represents the laminar Couette-solution.

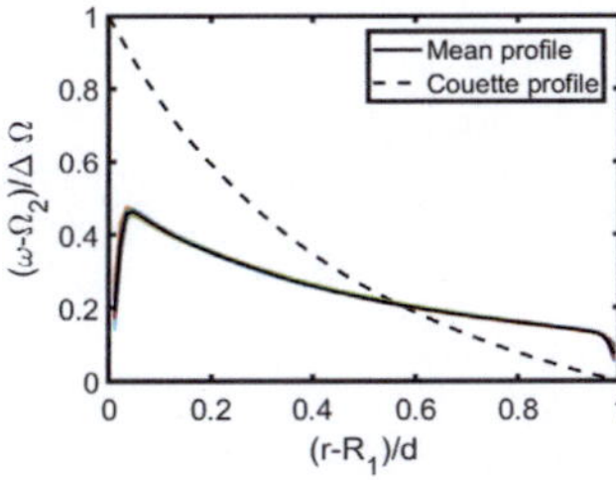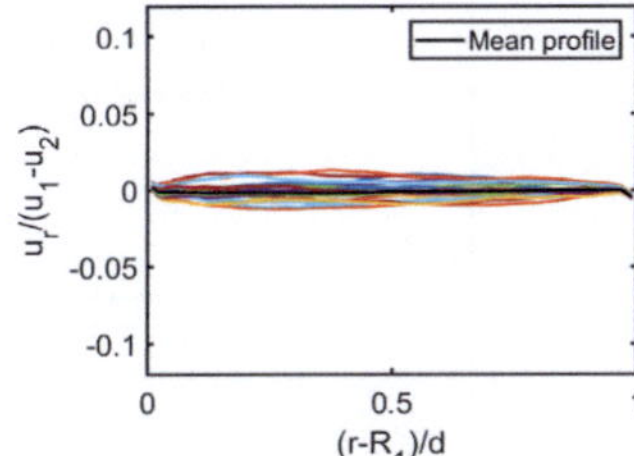

FIGURE B.38: Angular and radial velocity profiles averaged in time and azimuthal coordinate for $Re_S = 100000$, $\mu = 0$. In colour the different axial positions are depicted, the black solid line represents the axial mean profile and the dashed line in angular velocity profile represents the laminar Couette-solution.

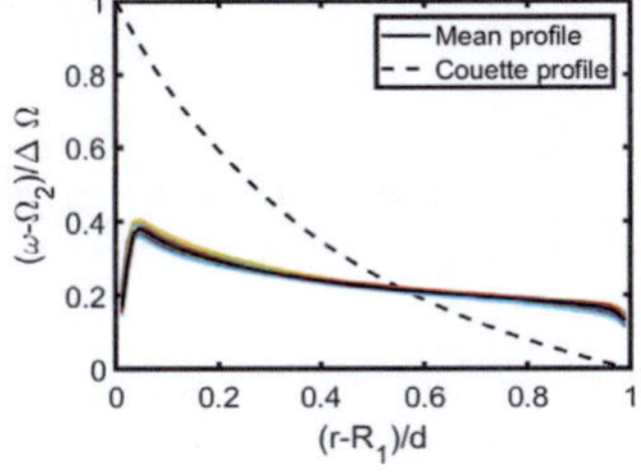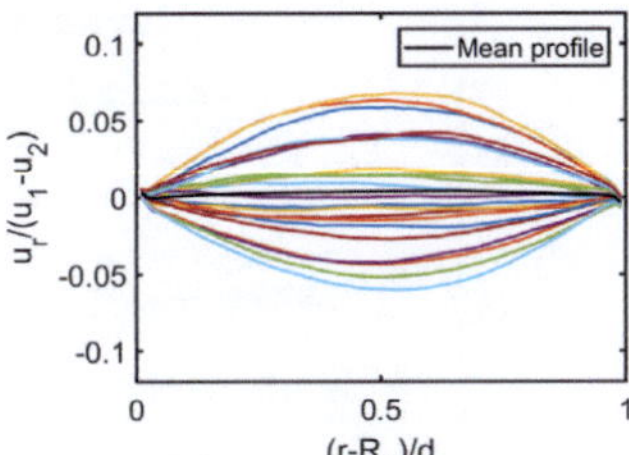

FIGURE B.39: Angular and radial velocity profiles averaged in time and azimuthal coordinate for $Re_S = 100000$, $\mu = -0.10$. In colour the different axial positions are depicted, the black solid line represents the axial mean profile and the dashed line in angular velocity profile represents the laminar Couette-solution.

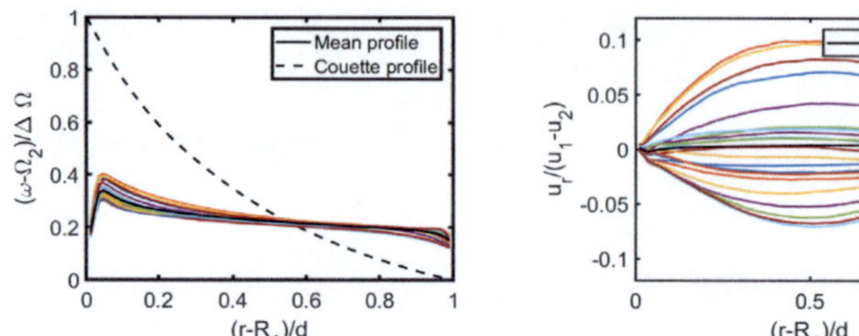

FIGURE B.40: Angular and radial velocity profiles averaged in time and azimuthal coordinate for $Re_S = 100000$, $\mu = -0.15$. In colour the different axial positions are depicted, the black solid line represents the axial mean profile and the dashed line in angular velocity profile represents the laminar Couette-solution.

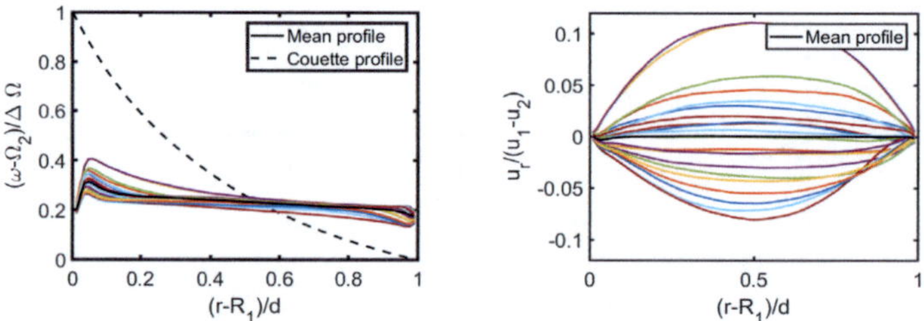

FIGURE B.41: Angular and radial velocity profiles averaged in time and azimuthal coordinate for $Re_S = 100000$, $\mu = -0.20$. In colour the different axial positions are depicted, the black solid line represents the axial mean profile and the dashed line in angular velocity profile represents the laminar Couette-solution.

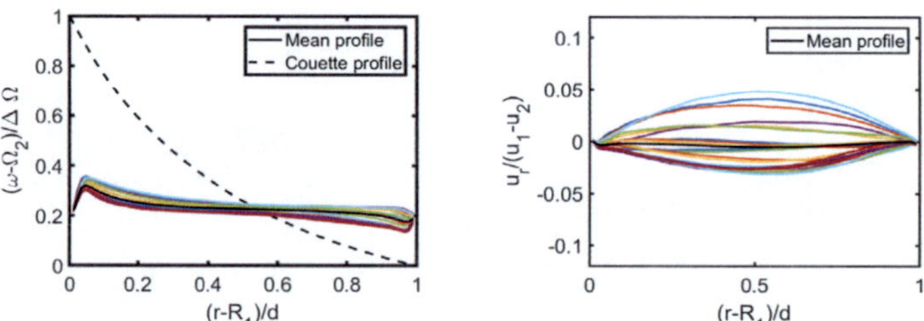

FIGURE B.42: Angular and radial velocity profiles averaged in time and azimuthal coordinate for $Re_S = 100000$, $\mu = -0.25$. In colour the different axial positions are depicted, the black solid line represents the axial mean profile and the dashed line in angular velocity profile represents the laminar Couette-solution.

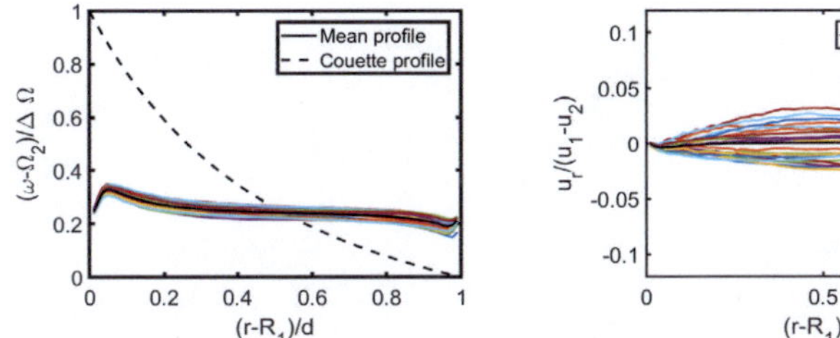

FIGURE B.43: Angular and radial velocity profiles averaged in time and azimuthal coordinate for $Re_S = 100000$, $\mu = -0.30$. In colour the different axial positions are depicted, the black solid line represents the axial mean profile and the dashed line in angular velocity profile represents the laminar Couette-solution.

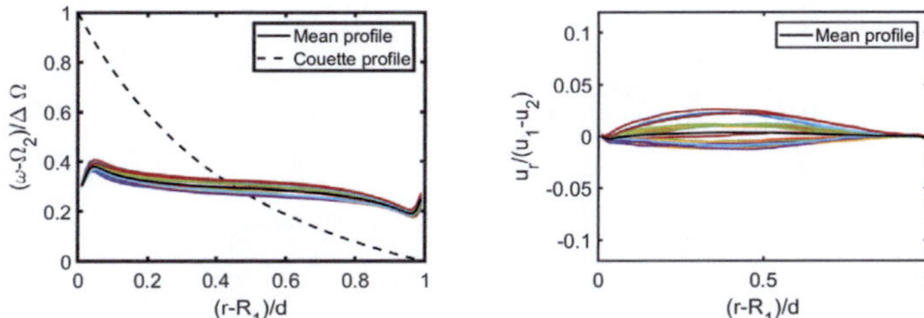

FIGURE B.44: Angular and radial velocity profiles averaged in time and azimuthal coordinate for $Re_S = 100000$, $\mu = -0.40$. In colour the different axial positions are depicted, the black solid line represents the axial mean profile and the dashed line in angular velocity profile represents the laminar Couette-solution.

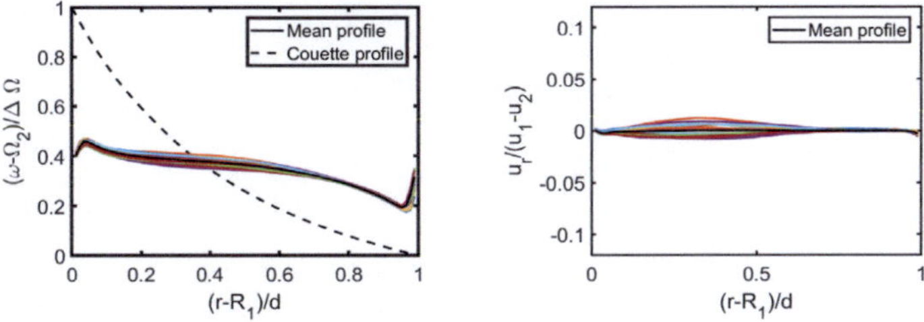

FIGURE B.45: Angular and radial velocity profiles averaged in time and azimuthal coordinate for $Re_S = 100000$, $\mu = -0.60$. In colour the different axial positions are depicted, the black solid line represents the axial mean profile and the dashed line in angular velocity profile represents the laminar Couette-solution.

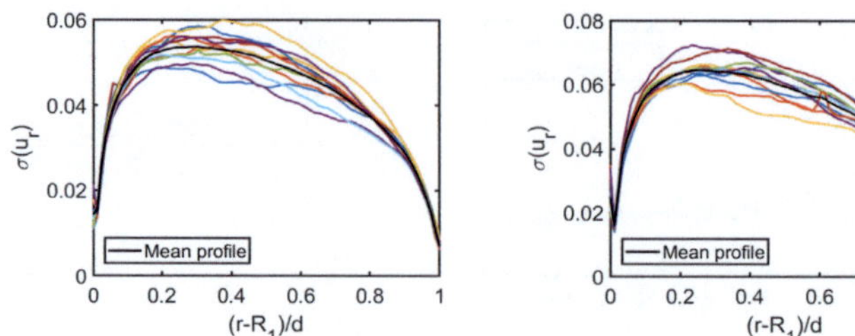

FIGURE B.46: Profiles of the standard deviation of the radial velocity component computed in time and azimuthal coordinate for $Re_S = 40000$ (left) and $Re_S = 50000$ (right) $\mu = 0$. In colour the different axial positions are depicted, the black solid line represents the axial mean profile.

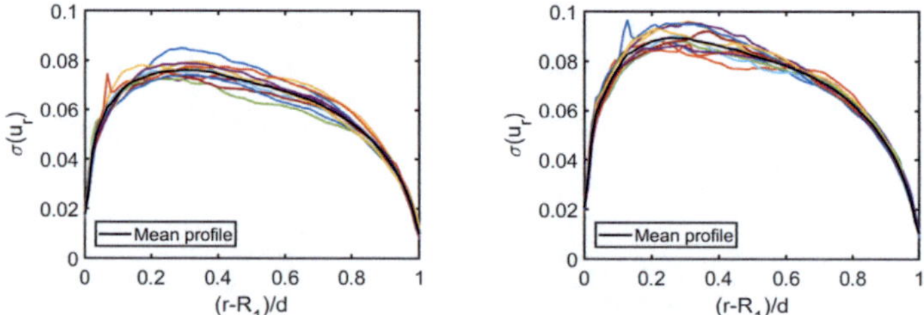

FIGURE B.47: Profiles of the standard deviation of the radial velocity component computed in time and azimuthal coordinate for $Re_S = 60000$ (left) and $Re_S = 70000$ (right) $\mu = 0$. In colour the different axial positions are depicted, the black solid line represents the axial mean profile.

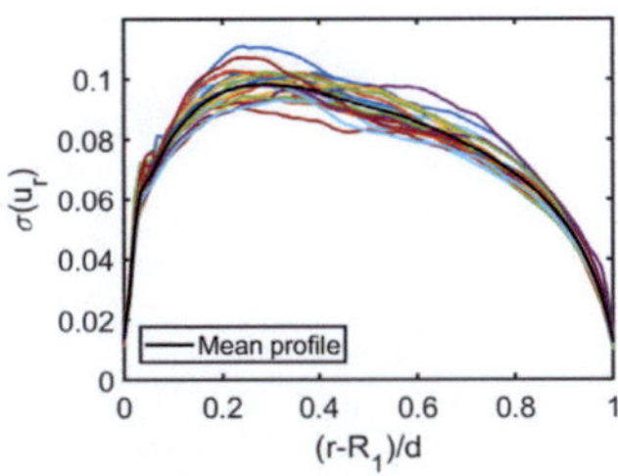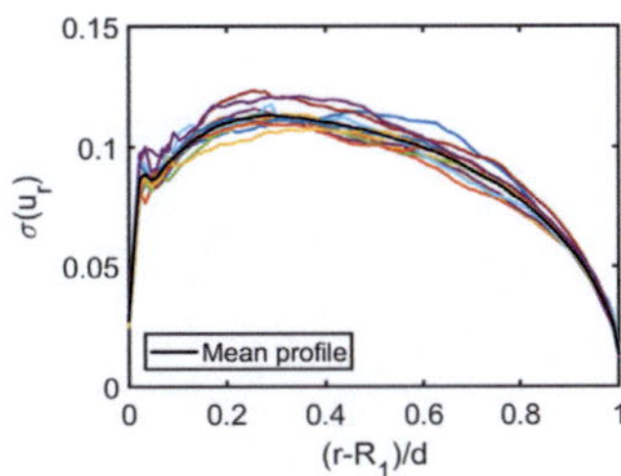

FIGURE B.48: Profiles of the standard deviation of the radial velocity component computed in time and azimuthal coordinate for $Re_S = 80000$ (left) and $Re_S = 90000$ (right) $\mu = 0$. In colour the different axial positions are depicted, the black solid line represents the axial mean profile.

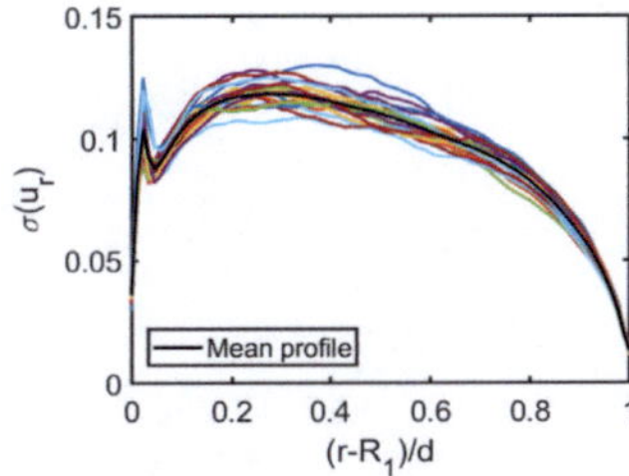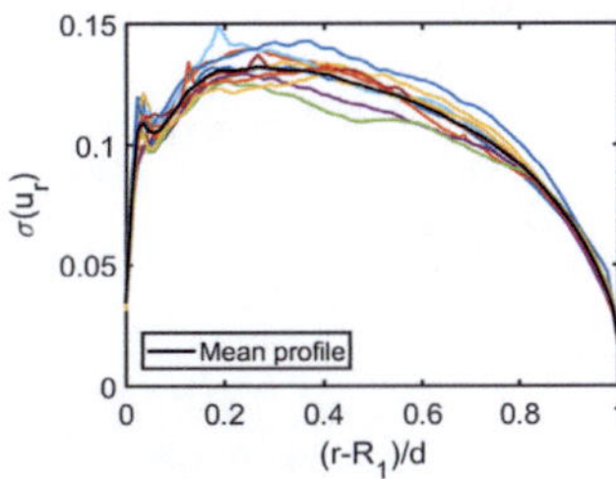

FIGURE B.49: Profiles of the standard deviation of the radial velocity component computed in time and azimuthal coordinate for $Re_S = 100000$ (left) and $Re_S = 110000$ (right) $\mu = 0$. In colour the different axial positions are depicted, the black solid line represents the axial mean profile.

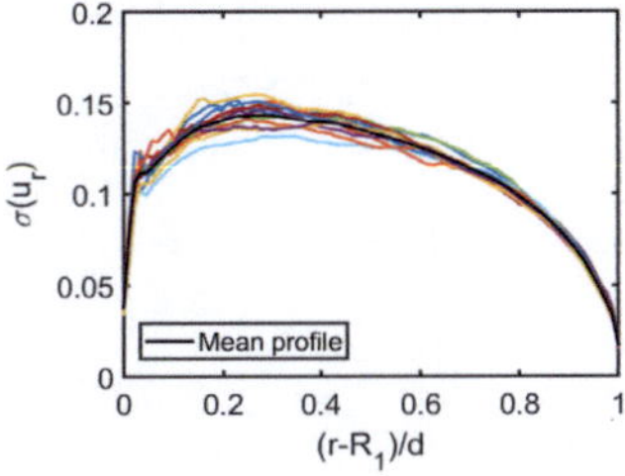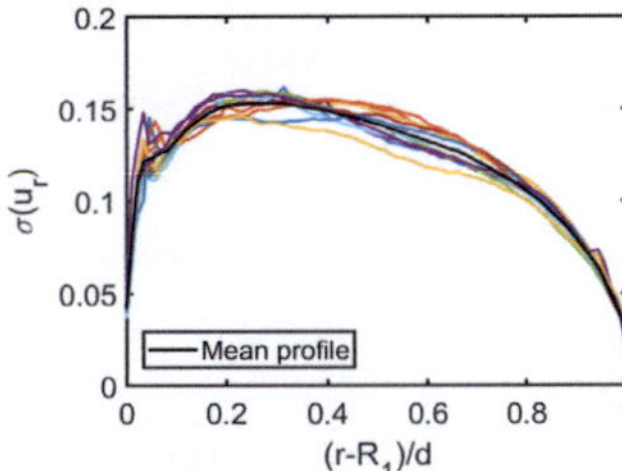

FIGURE B.50: Profiles of the standard deviation of the radial velocity component computed in time and azimuthal coordinate for $Re_S = 120000$ (left) and $Re_S = 130000$ (right) $\mu = 0$. In colour the different axial positions are depicted, the black solid line represents the axial mean profile.

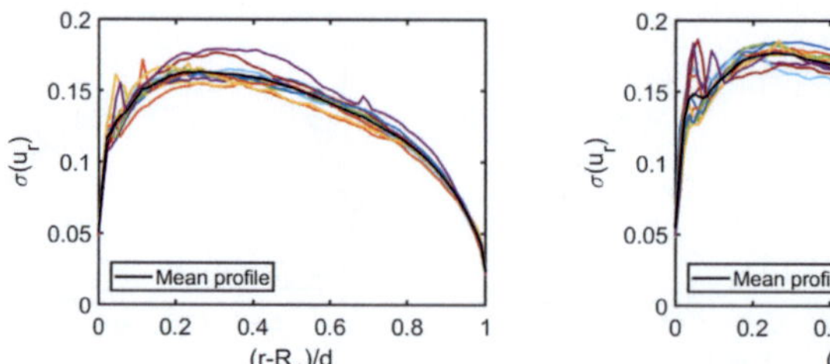
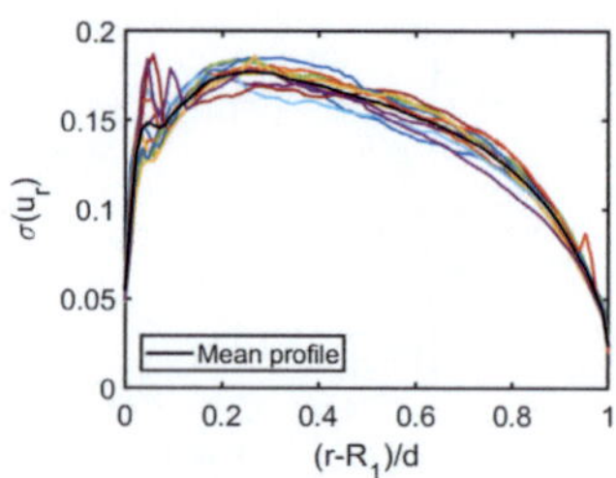

FIGURE B.51: Profiles of the standard deviation of the radial velocity component computed in time and azimuthal coordinate for $Re_S = 140000$ (left) and $Re_S = 150000$ (right) $\mu = 0$. In colour the different axial positions are depicted, the black solid line represents the axial mean profile.

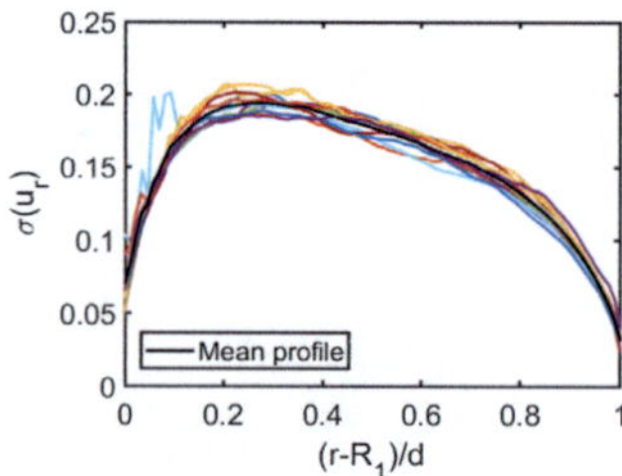
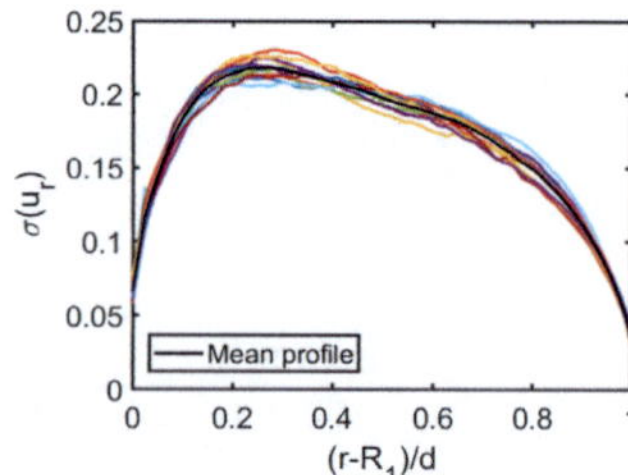

FIGURE B.52: Profiles of the standard deviation of the radial velocity component computed in time and azimuthal coordinate for $Re_S = 165000$ (left) and $Re_S = 192000$ (right) $\mu = 0$. In colour the different axial positions are depicted, the black solid line represents the axial mean profile.

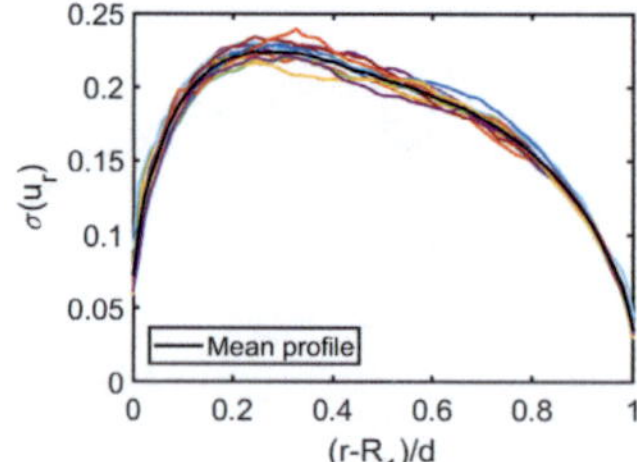
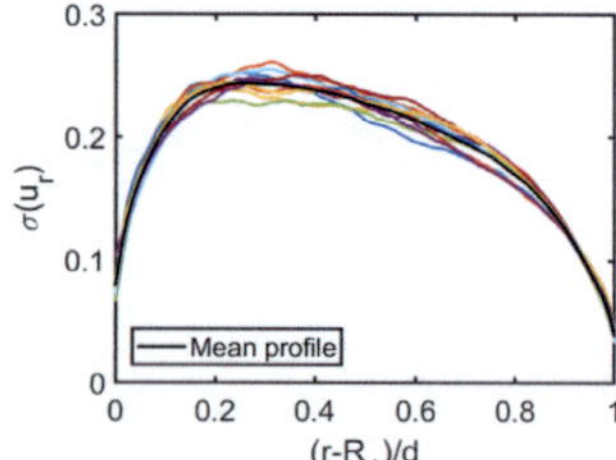

FIGURE B.53: Profiles of the standard deviation of the radial velocity component computed in time and azimuthal coordinate for $Re_S = 201000$ (left) and $Re_S = 219000$ (right) $\mu = 0$. In colour the different axial positions are depicted, the black solid line represents the axial mean profile.

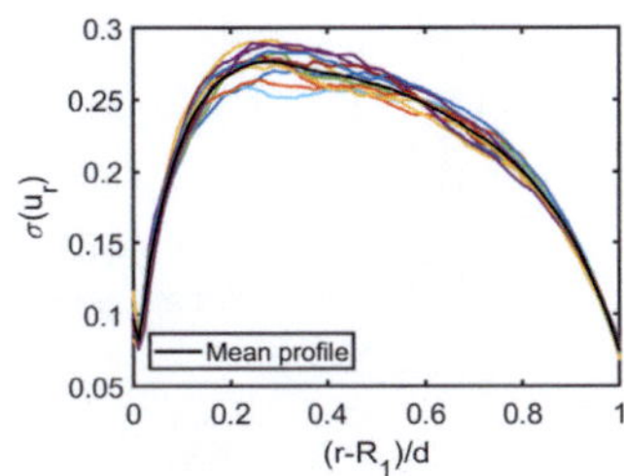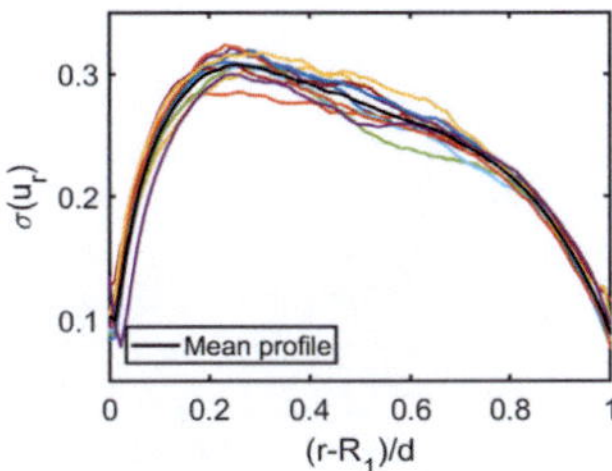

FIGURE B.54: Profiles of the standard deviation of the radial velocity component computed in time and azimuthal coordinate for $Re_S = 245000$ (left) and $Re_S = 270000$ (right) $\mu = 0$. In colour the different axial positions are depicted, the black solid line represents the axial mean profile.

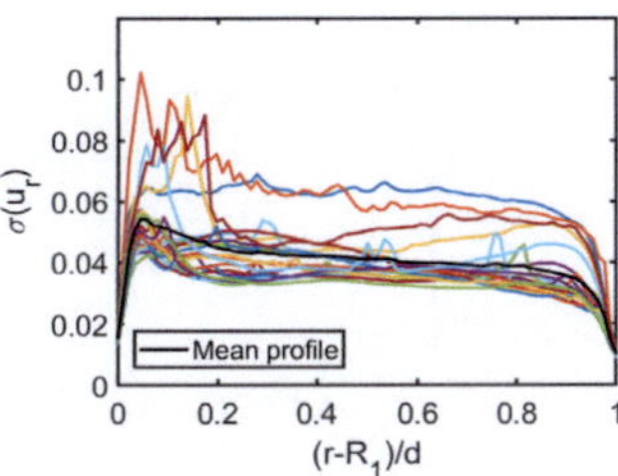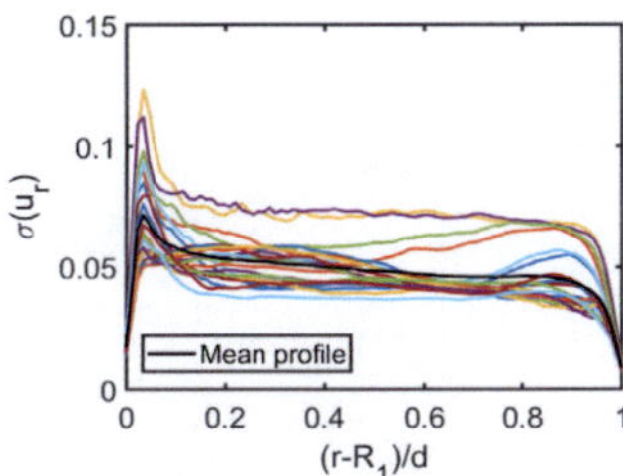

FIGURE B.55: Profiles of the standard deviation of the radial velocity component computed in time and azimuthal coordinate for $Re_S = 80000$ (left) and $Re_S = 100000$ (right) $\mu = -0.20$. In colour the different axial positions are depicted, the black solid line represents the axial mean profile.

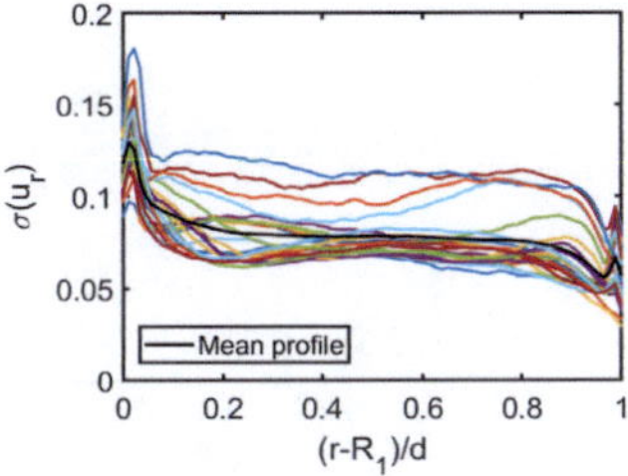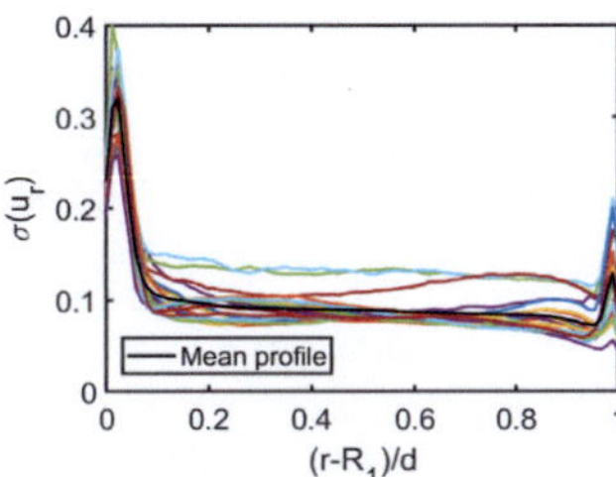

FIGURE B.56: Profiles of the standard deviation of the radial velocity component computed in time and azimuthal coordinate for $Re_S = 165000$ (left) and $Re_S = 192000$ (right) $\mu = -0.20$. In colour the different axial positions are depicted, the black solid line represents the axial mean profile.

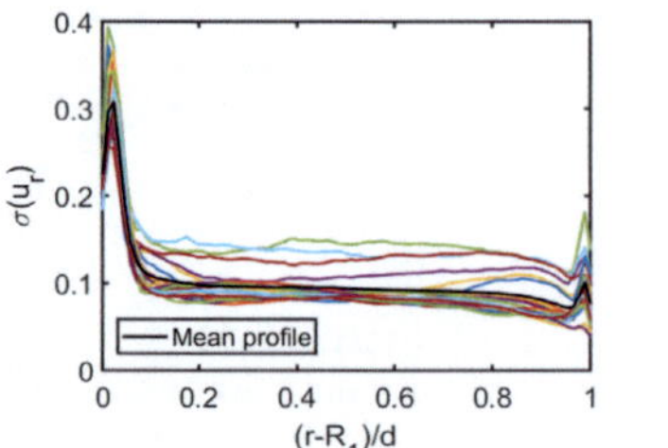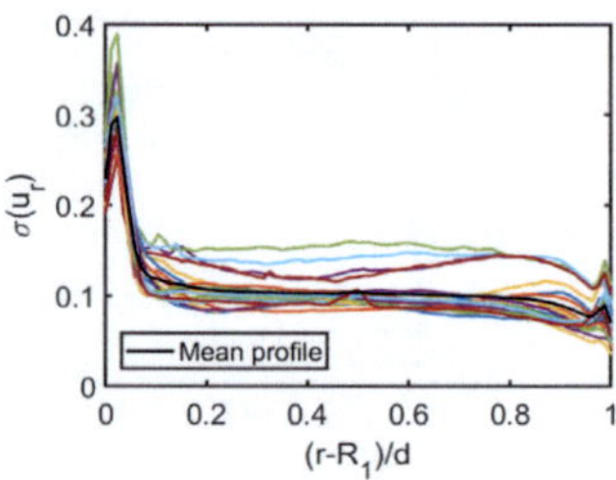

FIGURE B.57: Profiles of the standard deviation of the radial velocity component computed in time and azimuthal coordinate for $Re_S = 201000$ (left) and $Re_S = 219000$ (right) $\mu = -0.20$. In colour the different axial positions are depicted, the black solid line represents the axial mean profile.

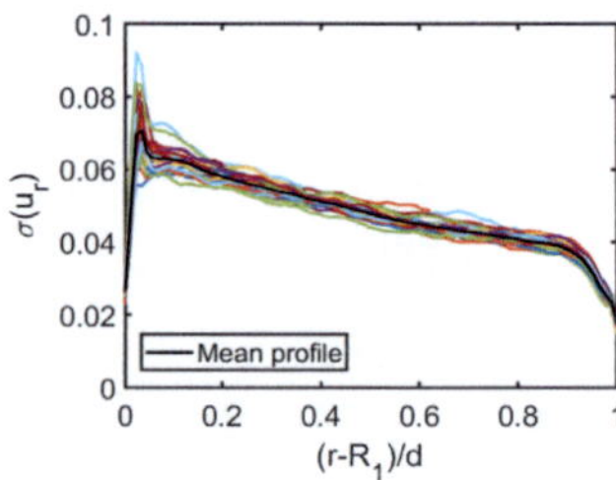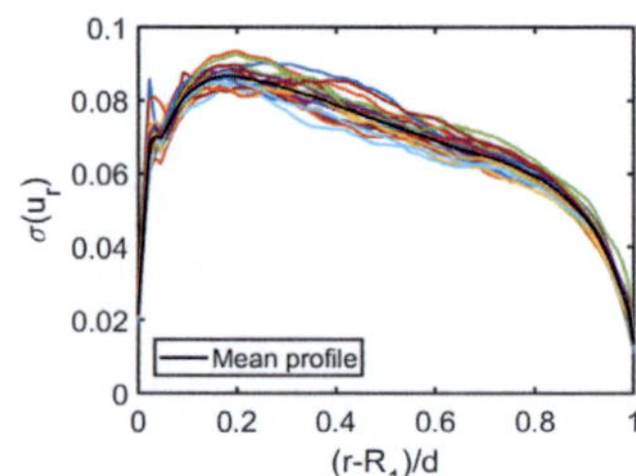

FIGURE B.58: Profiles of the standard deviation of the radial velocity component computed in time and azimuthal coordinate for $Re_S = 80000$ $\mu = +0.20$ (left) and $\mu = +0.10$ (right). In colour the different axial positions are depicted, the black solid line represents the axial mean profile.

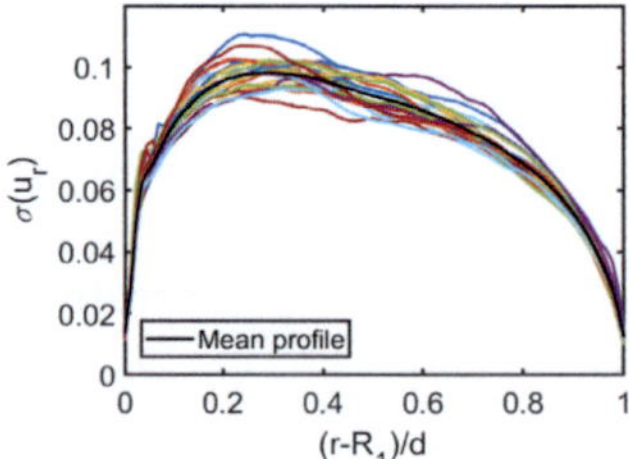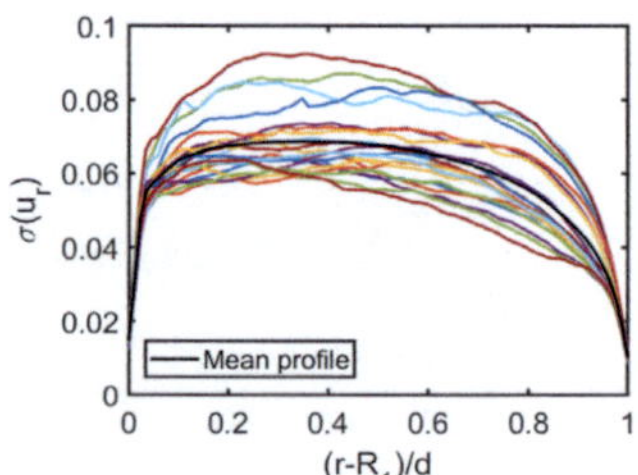

FIGURE B.59: Profiles of the standard deviation of the radial velocity component computed in time and azimuthal coordinate for $Re_S = 80000$ $\mu = 0$ (left) and $\mu = -0.10$ (right). In colour the different axial positions are depicted, the black solid line represents the axial mean profile.

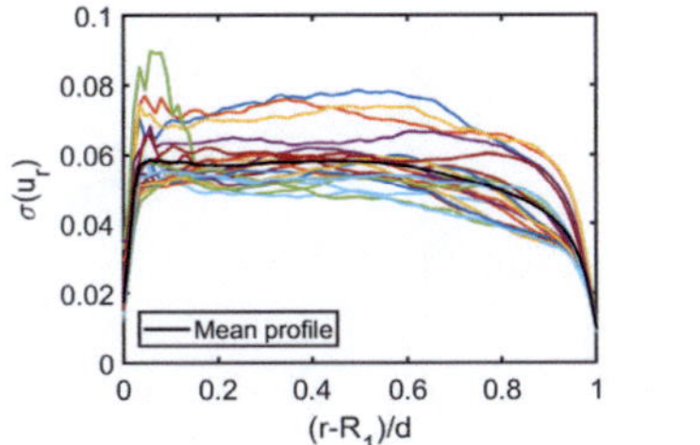 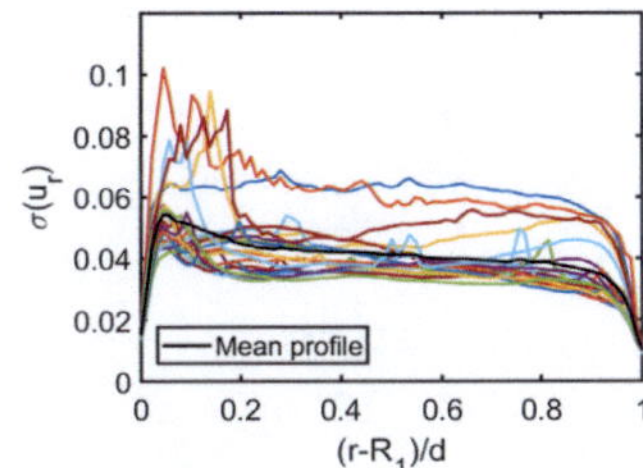

FIGURE B.60: Profiles of the standard deviation of the radial velocity component computed in time and azimuthal coordinate for $Re_S = 80000$ $\mu = -0.15$ (left) and $\mu = -0.20$ (right). In colour the different axial positions are depicted, the black solid line represents the axial mean profile.

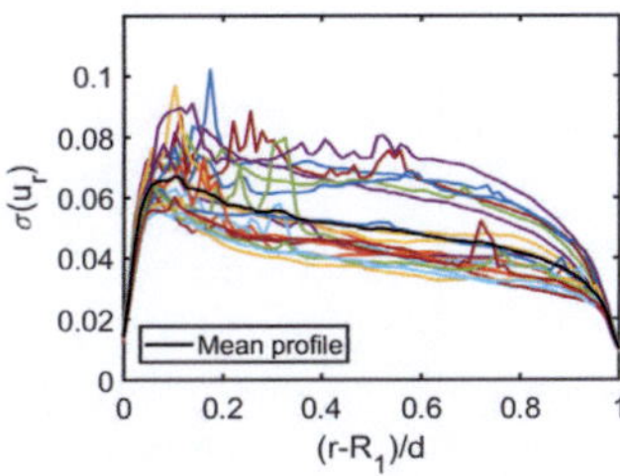 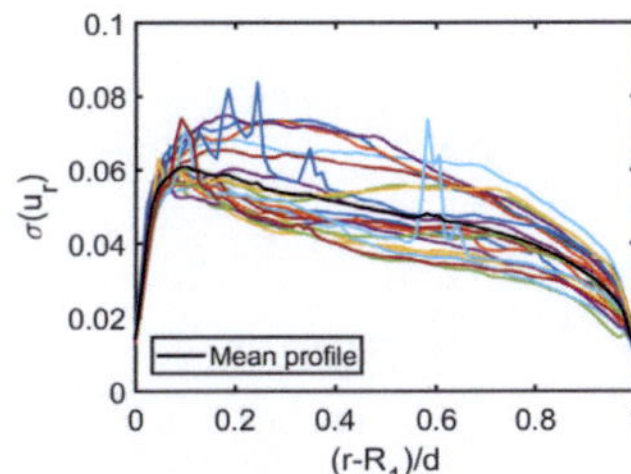

FIGURE B.61: Profiles of the standard deviation of the radial velocity component computed in time and azimuthal coordinate for $Re_S = 80000$ $\mu = -0.25$ (left) and $\mu = -0.30$ (right). In colour the different axial positions are depicted, the black solid line represents the axial mean profile.

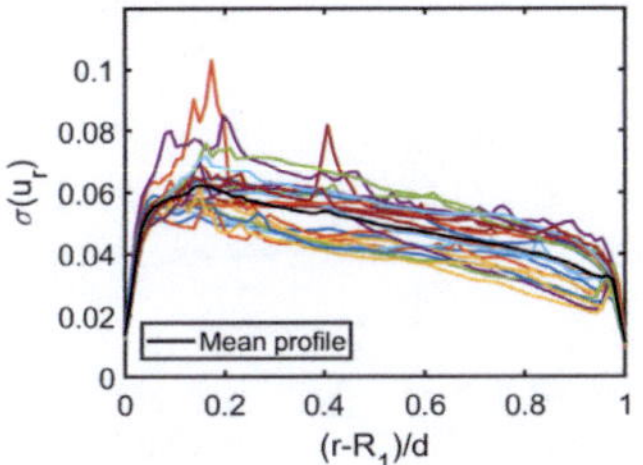 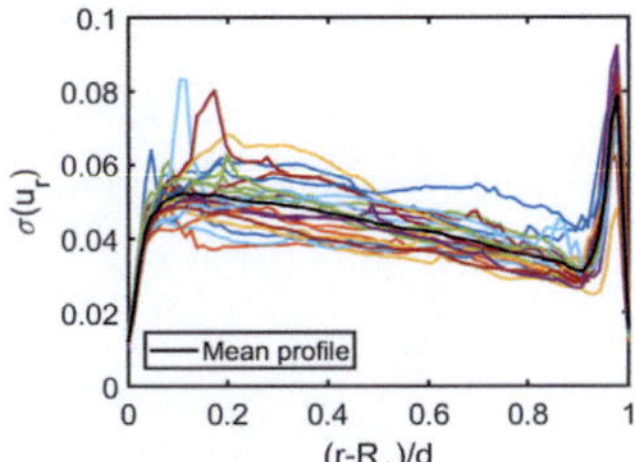

FIGURE B.62: Profiles of the standard deviation of the radial velocity component computed in time and azimuthal coordinate for $Re_S = 80000$ $\mu = -0.40$ (left) and $\mu = -0.60$ (right). In colour the different axial positions are depicted, the black solid line represents the axial mean profile.

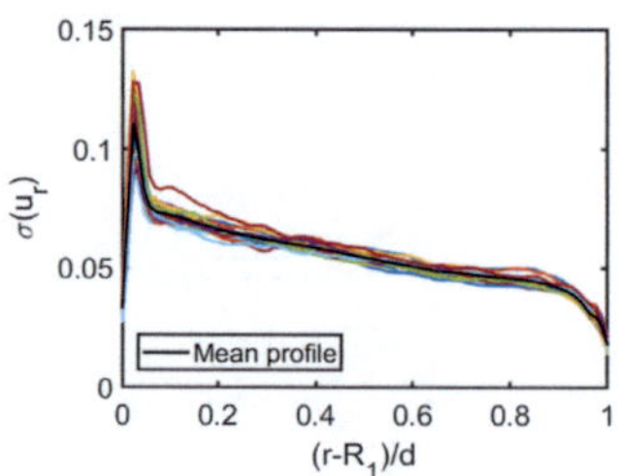 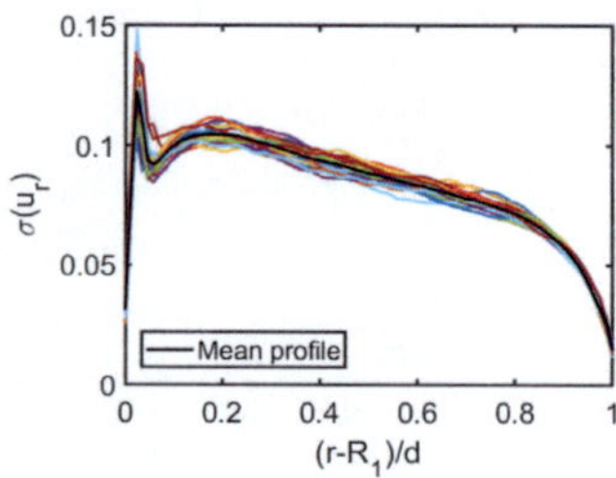

FIGURE B.63: Profiles of the standard deviation of the radial velocity component computed in time and azimuthal coordinate for $Re_S = 100000$ $\mu = +0.20$ (left) and $\mu = +0.10$ (right). In colour the different axial positions are depicted, the black solid line represents the axial mean profile.

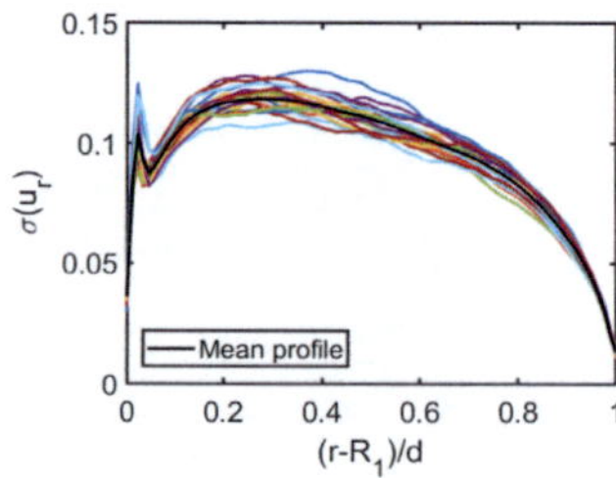 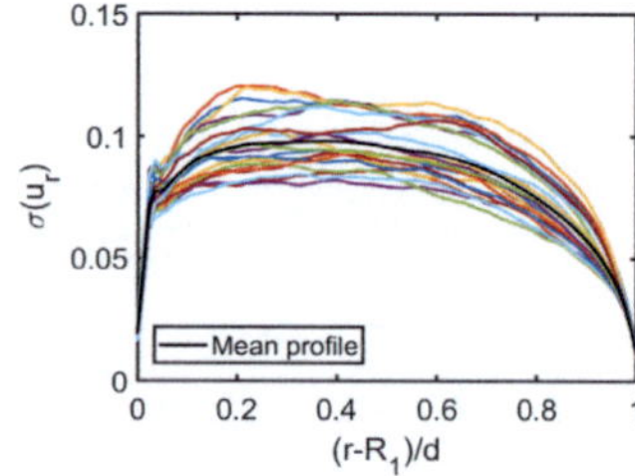

FIGURE B.64: Profiles of the standard deviation of the radial velocity component computed in time and azimuthal coordinate for $Re_S = 100000$ $\mu = 0$ (left) and $\mu = -0.10$ (right). In colour the different axial positions are depicted, the black solid line represents the axial mean profile.

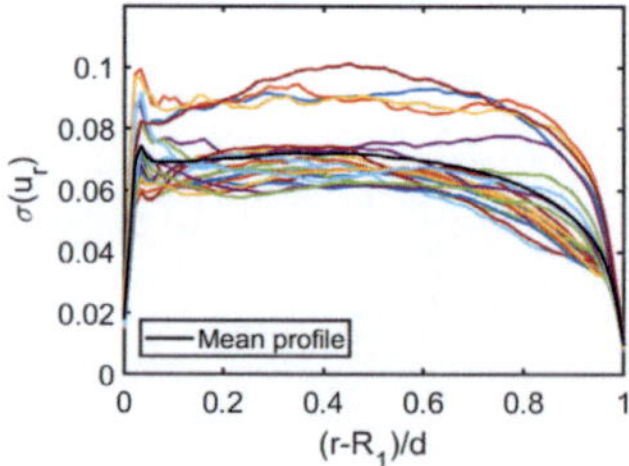 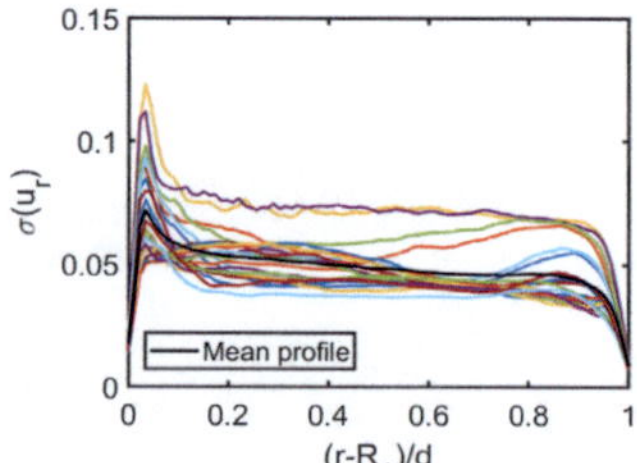

FIGURE B.65: Profiles of the standard deviation of the radial velocity component computed in time and azimuthal coordinate for $Re_S = 100000$ $\mu = -0.15$ (left) and $\mu = -0.20$ (right). In colour the different axial positions are depicted, the black solid line represents the axial mean profile.

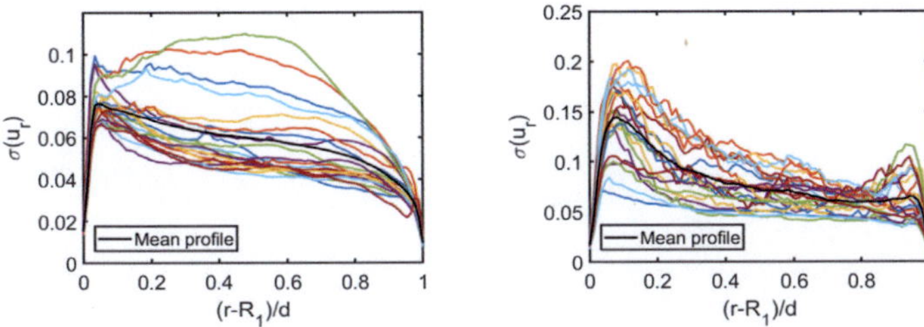

FIGURE B.66: Profiles of the standard deviation of the radial velocity component computed in time and azimuthal coordinate for $Re_S = 100000$ $\mu = -0.25$ (left) and $\mu = -0.30$ (right). In colour the different axial positions are depicted, the black solid line represents the axial mean profile.

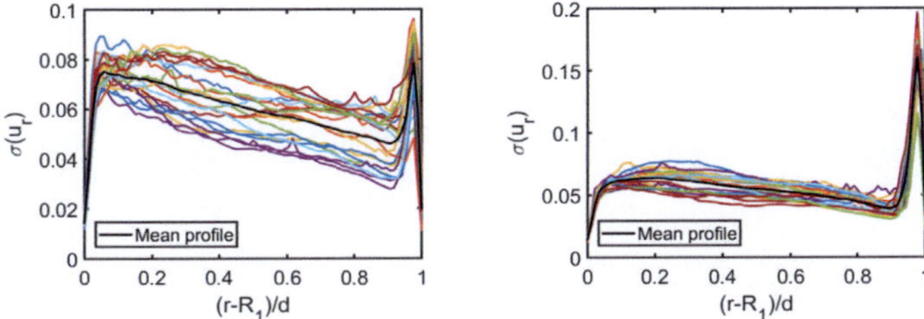

FIGURE B.67: Profiles of the standard deviation of the radial velocity component computed in time and azimuthal coordinate for $Re_S = 100000$ $\mu = -0.40$ (left) and $\mu = -0.60$ (right). In colour the different axial positions are depicted, the black solid line represents the axial mean profile.

Appendix C

Appendix C: PIV: Flow structures

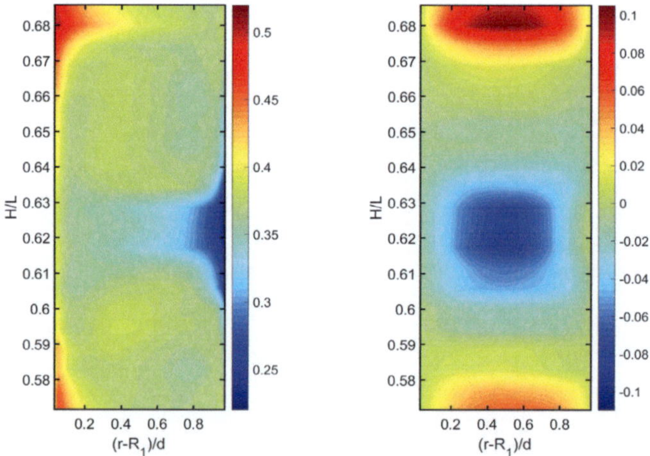

FIGURE C.1: Contours of the Angular (left) and radial (right) velocities averaged in time and azimuthal coordinate for $Re_S = 80000$, $\mu = -0.20$.

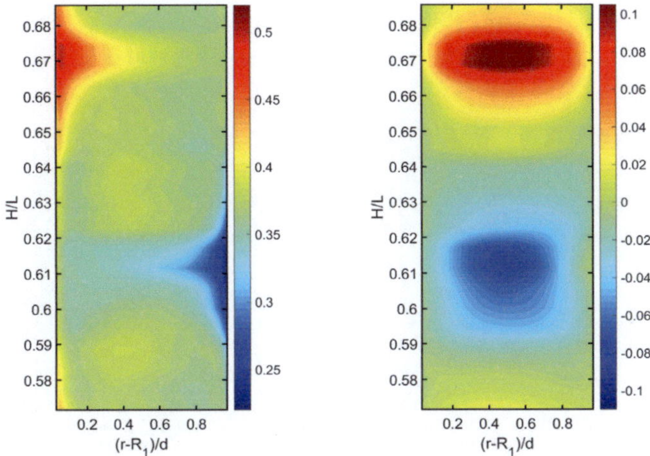

FIGURE C.2: Contours of the Angular (left) and radial (right) velocities averaged in time and azimuthal coordinate for $Re_S = 100000$, $\mu = -0.20$.

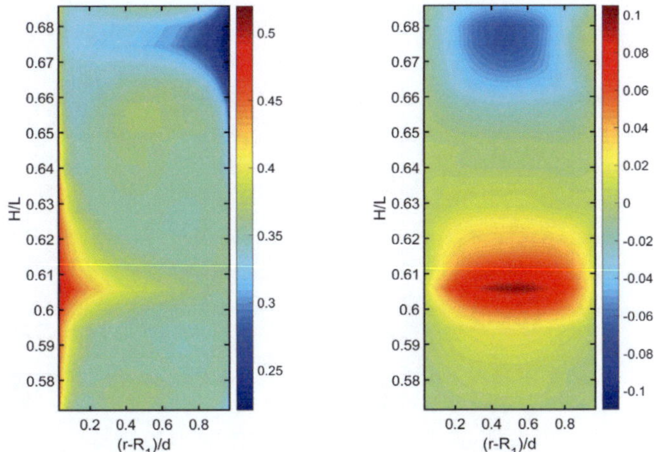

FIGURE C.3: Contours of the Angular (left) and radial (right) velocities averaged in time and azimuthal coordinate for $Re_S = 165000$, $\mu = -0.20$.

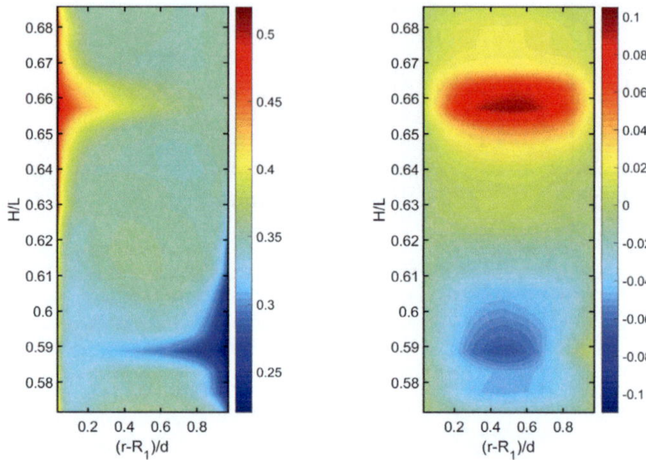

FIGURE C.4: Contours of the Angular (left) and radial (right) velocities averaged in time and azimuthal coordinate for $Re_S = 192000$, $\mu = -0.20$.

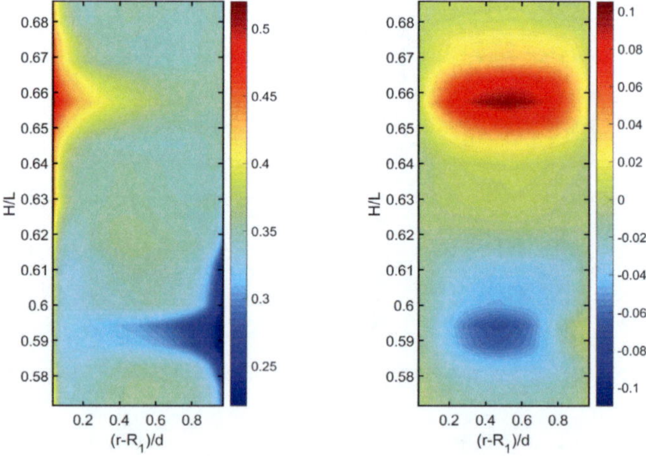

FIGURE C.5: Contours of the Angular (left) and radial (right) velocities averaged in time and azimuthal coordinate for $Re_S = 201000$, $\mu = -0.20$.

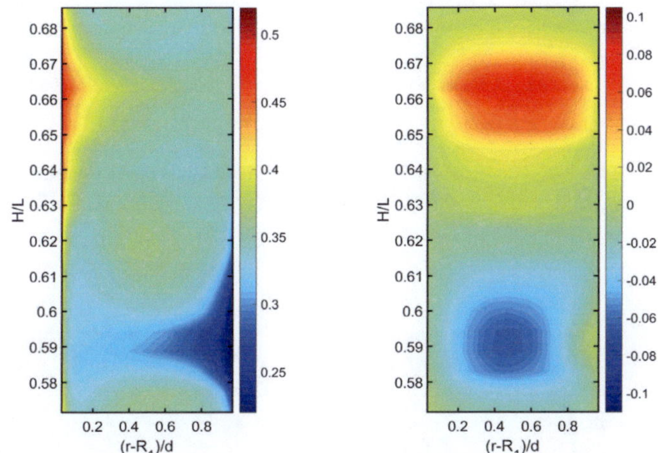

FIGURE C.6: Contours of the Angular (left) and radial (right) velocities averaged in time and azimuthal coordinate for $Re_S = 219000$, $\mu = -0.20$.

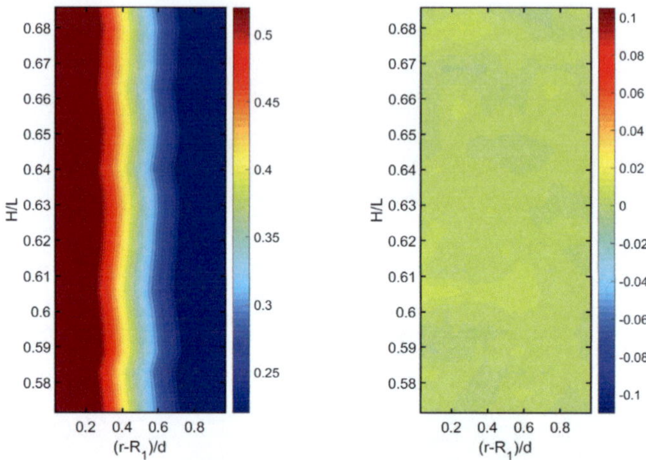

FIGURE C.7: Contours of the Angular (left) and radial (right) velocities averaged in time and azimuthal coordinate for $Re_S = 80000$, $\mu = +0.20$.

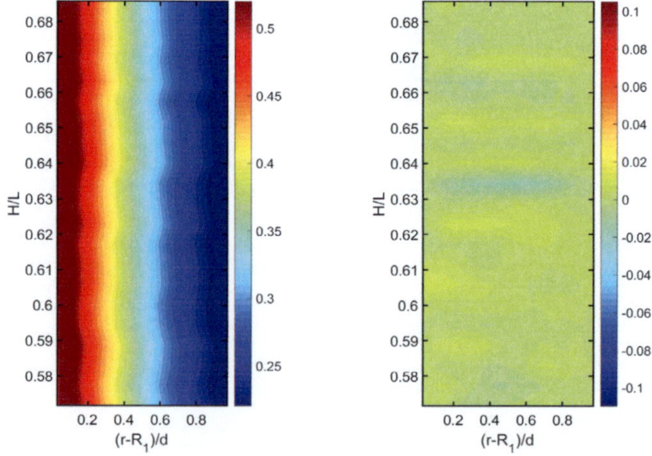

FIGURE C.8: Contours of the Angular (left) and radial (right) velocities averaged in time and azimuthal coordinate for $Re_S = 80000$, $\mu = +0.10$.

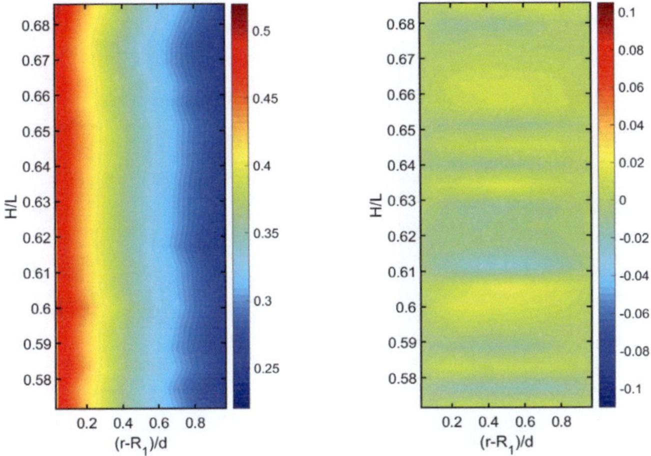

FIGURE C.9: Contours of the Angular (left) and radial (right) velocities averaged in time and azimuthal coordinate for $Re_S = 80000$, $\mu = 0$.

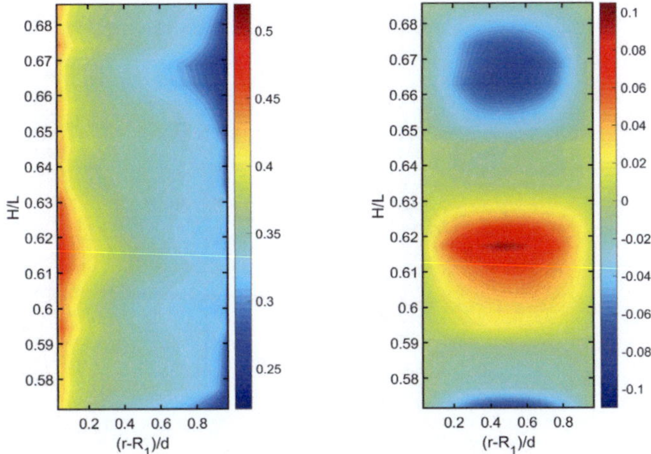

FIGURE C.10: Contours of the Angular (left) and radial (right) velocities averaged in time and azimuthal coordinate for $Re_S = 80000$, $\mu = -0.10$.

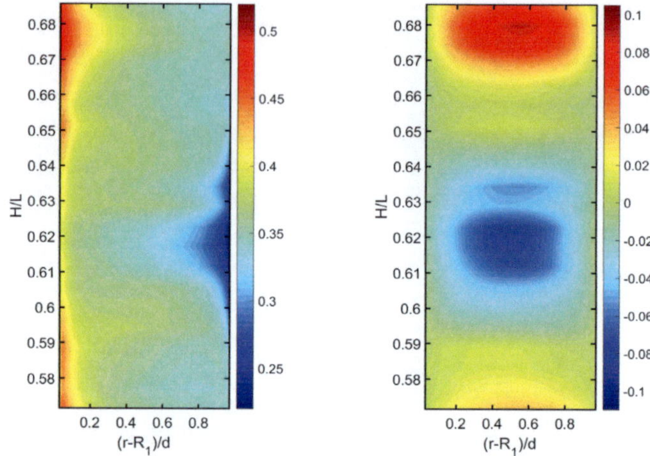

FIGURE C.11: Contours of the Angular (left) and radial (right) velocities averaged in time and azimuthal coordinate for $Re_S = 80000$, $\mu = -0.15$.

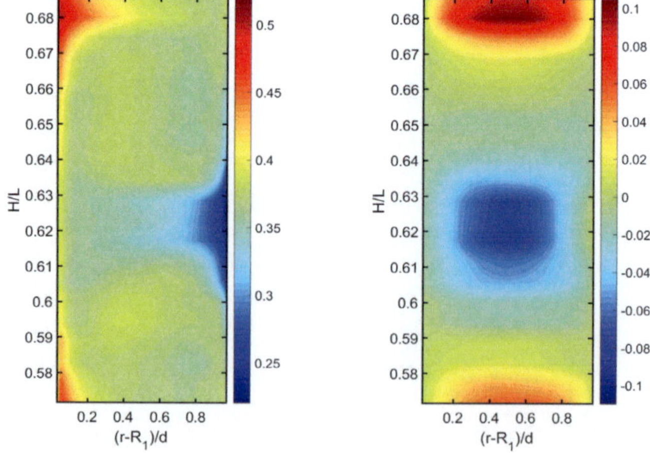

FIGURE C.12: Contours of the Angular (left) and radial (right) velocities averaged in time and azimuthal coordinate for $Re_S = 80000$, $\mu = -0.20$.

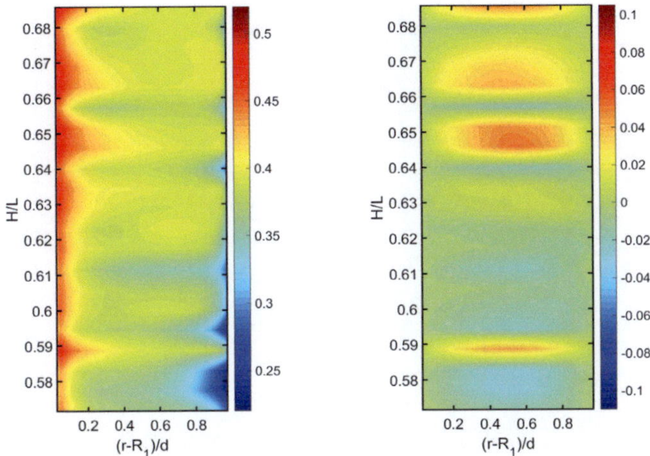

FIGURE C.13: Contours of the Angular (left) and radial (right) velocities averaged in time and azimuthal coordinate for $Re_S = 80000$, $\mu = -0.25$.

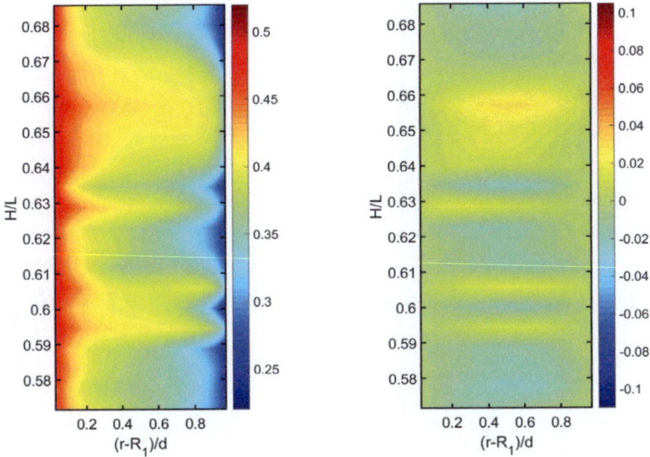

FIGURE C.14: Contours of the Angular (left) and radial (right) velocities averaged in time and azimuthal coordinate for $Re_S = 80000$, $\mu = -0.30$.

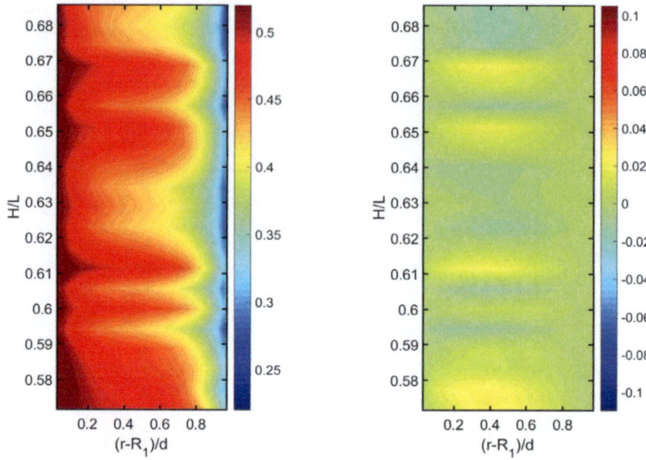

FIGURE C.15: Contours of the Angular (left) and radial (right) velocities averaged in time and azimuthal coordinate for $Re_S = 80000$, $\mu = -0.40$.

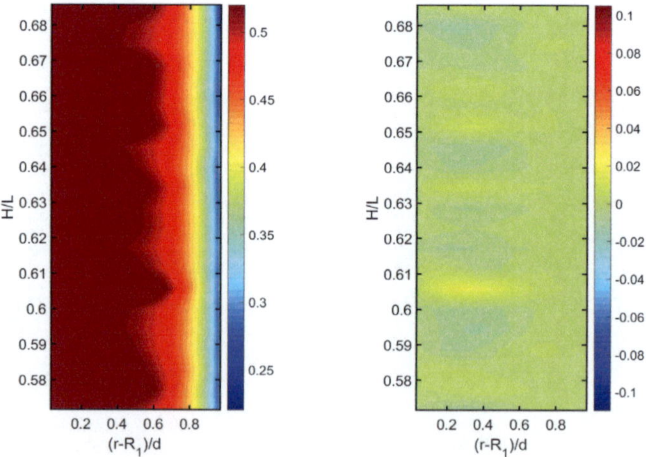

FIGURE C.16: Contours of the Angular (left) and radial (right) velocities averaged in time and azimuthal coordinate for $Re_S = 80000$, $\mu = -0.60$.

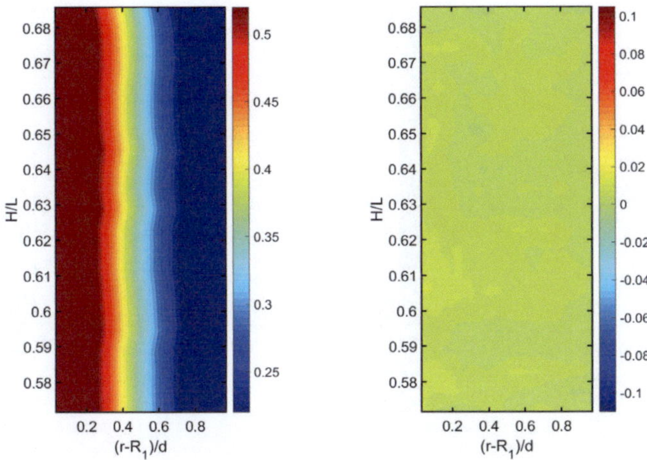

FIGURE C.17: Contours of the Angular (left) and radial (right) velocities averaged in time and azimuthal coordinate for $Re_S = 100000$, $\mu = +0.20$.

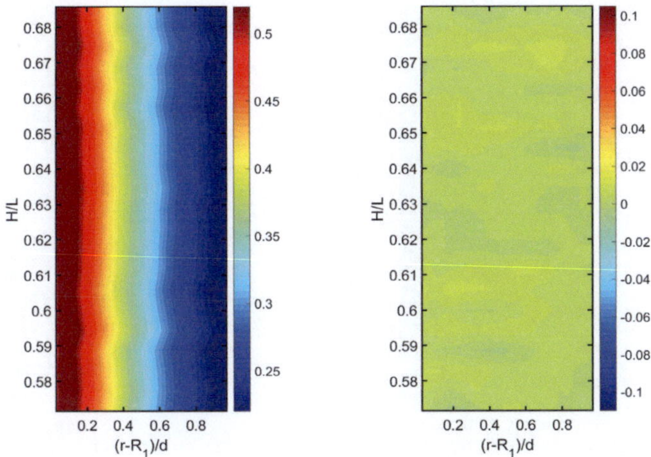

FIGURE C.18: Contours of the Angular (left) and radial (right) velocities averaged in time and azimuthal coordinate for $Re_S = 100000$, $\mu = +0.10$.

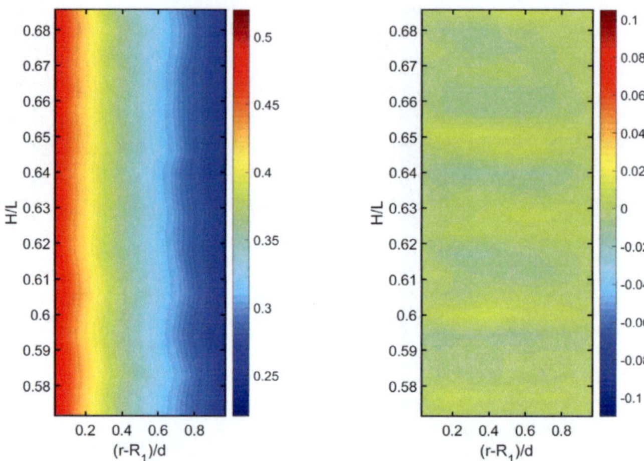

FIGURE C.19: Contours of the Angular (left) and radial (right) velocities averaged in time and azimuthal coordinate for $Re_S = 100000$, $\mu = 0$.

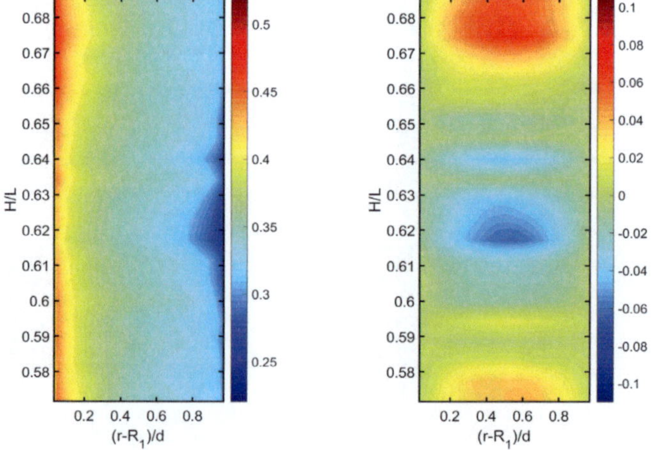

FIGURE C.20: Contours of the Angular (left) and radial (right) velocities averaged in time and azimuthal coordinate for $Re_S = 100000$, $\mu = -0.10$.

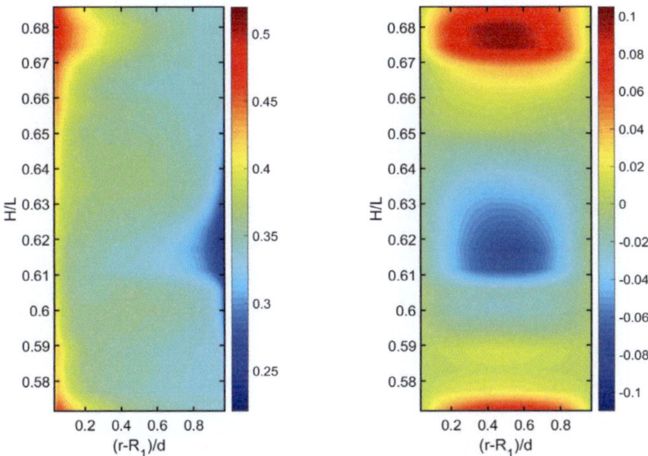

FIGURE C.21: Contours of the Angular (left) and radial (right) velocities averaged in time and azimuthal coordinate for $Re_S = 100000$, $\mu = -0.15$.

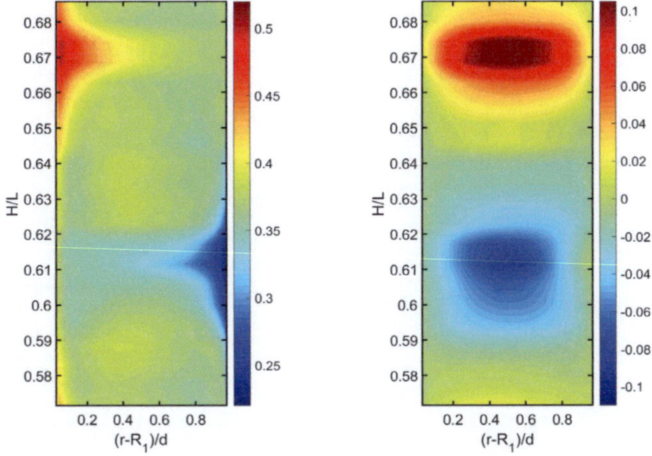

FIGURE C.22: Contours of the Angular (left) and radial (right) velocities averaged in time and azimuthal coordinate for $Re_S = 100000$, $\mu = -0.20$.

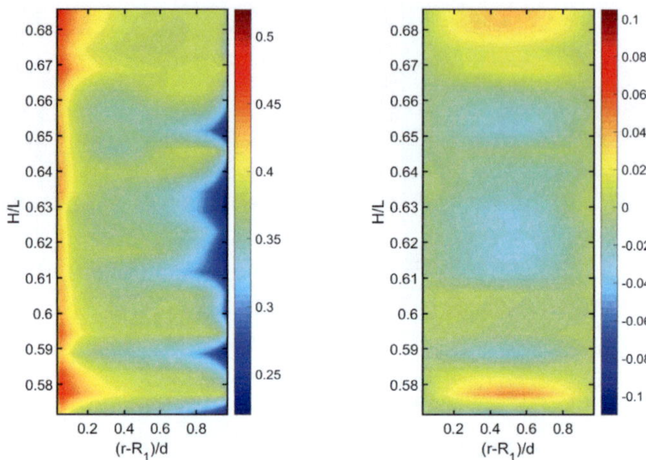

FIGURE C.23: Contours of the Angular (left) and radial (right) velocities averaged in time and azimuthal coordinate for $Re_S = 100000$, $\mu = -0.25$.

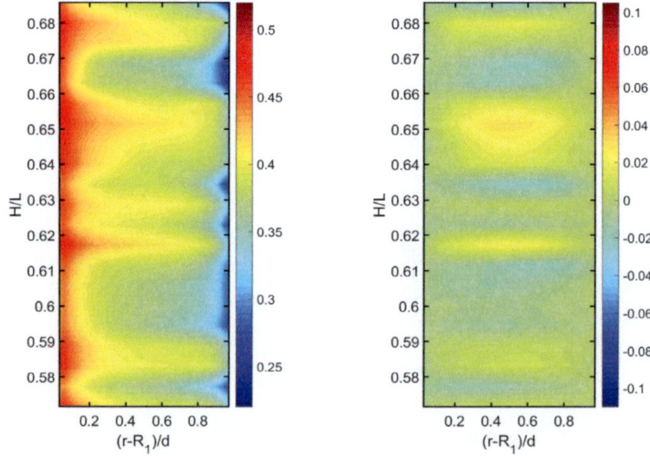

FIGURE C.24: Contours of the Angular (left) and radial (right) velocities averaged in time and azimuthal coordinate for $Re_S = 100000$, $\mu = -0.30$.

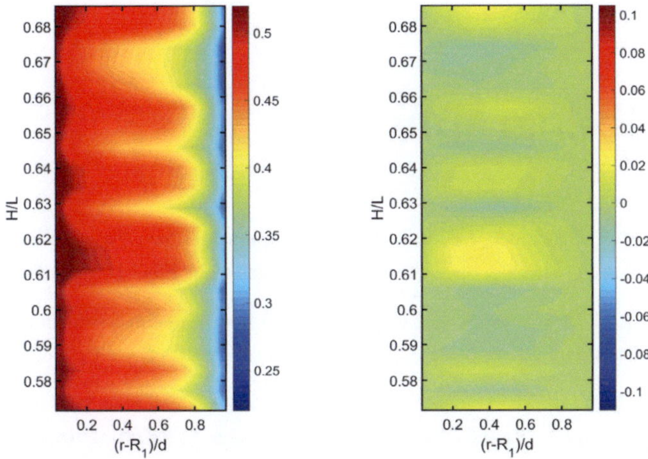

FIGURE C.25: Contours of the Angular (left) and radial (right) velocities averaged in time and azimuthal coordinate for $Re_S = 100000$, $\mu = -0.40$.

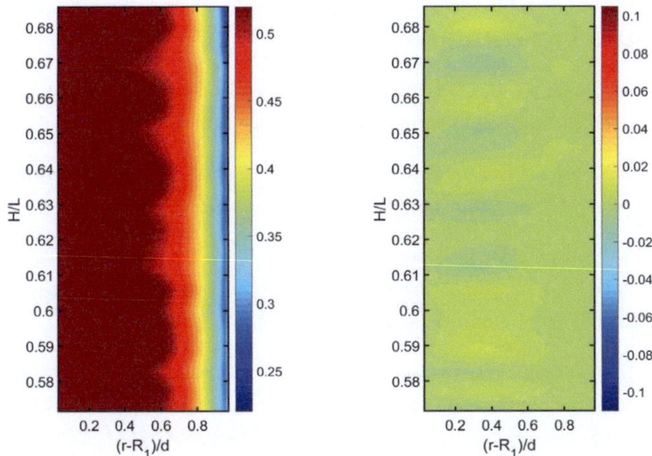

FIGURE C.26: Contours of the Angular (left) and radial (right) velocities averaged in time and azimuthal coordinate for $Re_S = 100000$, $\mu = -0.60$.

Bibliography

[1] C. D. Andereck, S.S. Liu, and H. L. Swinney. Flow regimes in a circular couette system with independently rotating cylinders. *J. Fluid Mech.*, 164:155–183, 1986.

[2] H. J. Brauckmann and B. Eckhardt. Direct numerical simulations of local and global torque in Taylor-Couette flow up to re=30.000. *JFM*, 718:398–427, 2013.

[3] H. J. Brauckmann and B. Eckhardt. Intermittent boundary layers and torque maxima in Taylor-Couette flow. *Phys. Rev. E*, 87:033004, 2013b.

[4] H. J. Brauckmann, M. Salewsky, and B. Eckhardt. Momentum transport in Taylor-Couette flow with vanishing curvature. *J. Fluid Mech.*, 790:419–452, 2016.

[5] M. Burin, E. Schartman, and H. Ji. Local measurements of turbulent angular momentum transport in circular Couette flow. *Exp. Fluids*, 48:763–769, 2010.

[6] S. Chandrasekhar. Hydrodynamic and Hydromagnetic Stability. *Clarendon Press, Oxford*, page 1st ed., 1961.

[7] A. Chouippe, E. Climent, D. Legendre, and C. Gabillet. Numerical simulation of bubble dispersion in turbulent Taylor-Couette flow. *Phys. Fluids*, 26:043304, 2014.

[8] S. Dong. Turbulent flow between counter-rotating concentric cylinders: a direct numerical simulation study. *J. Fluid Mech.*, 615:371–99, 2008.

[9] B. Dubrulle, O. Dauchot, F. Daviaud, P.-Y. Longaretti, D. Richard, and J.-P. Zahn. Stability and turbulent transport in Taylor-Couette flow from analysis of experimental data. *Phys. o Fluids*, 17:095103, 2005.

[10] B. Eckhardt, S. Grossmann, and D. Lohse. Fluxes and energy dissipation in thermal convection and shear flows. *EPL (Europhysics Letters)*, 78(2):24001, 2007. URL http://stacks.iop.org/0295-5075/78/i=2/a=24001.

[11] B. Eckhardt, S. Grossmann, and D. Lohse. Torque scaling in turbulent Taylor-Couette flow between independently rotating cylinders. *J. Fluid Mech.*, 581:221–250, 2007.

[12] A. Esser and S. Grossmann. Analytic expression for Taylor-Couette stability boundary. *Phys. Fluids*, 8:1814, 1996.

[13] A. Froitzheim, S. Merbold, and C. Egbers. Velocity profiles, flow structures and sclaing in a wide gap turbulent Taylor-Couette flow. *J. Fluid Mech.*, 831:330–357, 2017.

[14] S. Grossmann and D. Lohse. Scaling in thermal convection: a unifying theory. *J. Fluid Mech.*, 407:27–56, 2000.

[15] S. Grossmann and D. Lohse. Multiple scaling in the ultimate regime of thermal convection. *Phys. Fluids*, 23:045108, 2011.

[16] S. Grossmann, D. Lohse, and C. Sun. Velocity profiles in strongly turbulent taylor–couette flow. *Phys. Fluids*, 26:025114, 2014.

[17] S. G. Huisman, D.P.M. van Gils, S. Grossmann, C. Sun, and D. Lohse. Ultimate turbulent Taylor-Couette flow. *PRL*, 108:024501, 2012.

[18] S. G. Huisman, R. C. A. van der Veen, C. Sun, and D. Lohse. Multiple states in highly turbulent Taylor-Couette flow. *Nat. Commun.*, 5:3820, 2014.

[19] H. Ji, M. Burin, E. Schartman, and J. Goodman. Hydrodynamic turbulence cannot transport angular momentum effectively in astrophysical disks. *Nature*, 444:343, 2006.

[20] D. P. Lathrop, J. Fineberg, and H. L. Swinney. Transition to shear-driven turbulence in Couette-Taylor flow. *Phys. Rev. A*, 46:6390, 1992.

[21] G.S. Lewis and H.L. Swinney. Velocity structure functions, scaling, and transitions in high reynolds-number Couette-Taylor flow. *Phys. Rev. E*, 59:5, 1999.

[22] B. Martinez-Arias, J. Peixinho, O. Crumeyrolle, and I. Mutabazi. Effect of the number of vortices on the torque scaling in Taylor-Couette flow. *J. Fluid Mech.*, 748:756–767, 2014.

[23] I. Marusic, J. P. Monty, M. Hultmark, and A. J. Smits. On the logarithmic region in wall turbulence. *J. Fluid Mech.*, 716:043304, 2013.

[24] S. Merbold, H. J. Brauckmann, and C. Egbers. Torque measurements and numerical determination in differentially rotating wide gap Taylor-Couette flow. *Phys. Rev. E*, 87:023014, 2013.

[25] A. Meseguer, M. Avila, F. Mellibovsky, and F. Marques. Solenoidal spectral formulations for the computation of secondary flows in cylindrical and annular geometries. *EPJ ST*, 146 (1):249–259, 2007.

[26] H. Nobach and E. Bodenschatz. Limitations of accuracy in PIV due to individual variations of particle image intensities. *Exp. Fluids*, 47:27–38, 2009.

[27] R. Ostilla-Mónico, S. G. Huisman, T. J. G. Janninik, D. P. M. van Gils, R. Verzicco, S. Grossmann, and D. Lohse. Optimal Taylor-Couette flow: radius ratio dependence. *J. Fluid Mech.*, 747:1–29, 2014a.

[28] R. Ostilla-Mónico, E. P. van der Poel, R. Verzicco, S. Grossmann, and D. Lohse. Exploring the phase diagram of fully turbulent Taylor-Couette flow. *J. Fluid Mech.*, 761:1–26, 2014b.

[29] R. Ostilla-Mónico, R. Verzicco, S. Grossmann, and D. Lohse. Optimal Taylor-Couette flow: radius ratio dependence. *J. Fluid Mech.*, 788:95–117, 2016.

[30] M.S. Paoletti and D.P. Lathrop. Angular momentum transport in turbulent flow between independently rotating cylinders. *PRL*, 106:024501, 2011.

[31] M.S. Paoletti, D.P.M. van Gils, B. Dubrulle, C. Sun, D. Lohse, and D.P. Lathrop. Angular momentum transport and turbulence in laboratory models of Keplerian flows. *Astr. and Astrophys.*, 547:A64, 2012.

[32] M. Raffel, C. E. Willert, F. Scarano, C. J. Kähler, S. T. Wereley, and J. Kompenhans. *Particle Image Velocimetry, a practical guide*, volume 3. Edition. 2018. ISBN 978-3-319-68851-0.

[33] F. Ravelet, R. Delfos, and J. Westerweel. Influence of global rotation and reynolds number on the large-scale features of a turbulent Taylor-Couette flow. *Phys. of Fluids*, 22:055103, 2010.

[34] J. C. Rotta. *Turbulente Strömungen*, volume 1. Edition. 1972. ISBN 3-519-02316-4.

[35] G.I. Taylor. Stability of a viscous liquid contained between two rotating concentric cylinders. *Philos. Trans. R. Soc. London*, A 223:289–343, 1923.

[36] R. C. A. van der Veen, S. G. Huisman, S. Merbold, U. Harlander, C. Egbers, D. Lohse, and C. Sun. Taylor-Couette turbulence at radius ratio 0.5: scaling, flow structures and plumes. *J. Fluid Mech.*, 799:334–351, 2016.

[37] D. P. M. van Gils, S. Grossmann, C. Sun, and D. Lohse. Optimal Taylor-Couette turbulence. *J. Fluid Mech.*, 706:118, 2012.

[38] D.P.M. van Gils, S. G. Huisman, G.-W. Bruggert, C. Sun, and D. Lohse. Torque scaling in turbulent Taylor-Couette flow with co- and counter rotating cylinders. *PRL*, 106:024502, 2011.

[39] F. Wendt. Turbulente Stroemungen zwischen zwei rotierenden konaxialen Zylindern. *Ing. Arch.*, 4:577–595, 1933.